THE PHYSICS OF SELENIUM
AND TELLURIUM

THE PHYSICS OF SELENIUM AND TELLURIUM

Proceedings of the International Symposium
held at Montreal, Canada
October 12–13, 1967
and sponsored by the
Selenium–Tellurium Development Association, Inc.
345 East 47th Street, New York, N.Y. 10017

Edited by

W. CHARLES COOPER

Head, Research Division, Noranda Research Centre,
240 Hymus Boulevard, Pointe Claire,
Quebec, Canada

PERGAMON PRESS

OXFORD · LONDON · EDINBURGH · NEW YORK
TORONTO · SYDNEY · PARIS · BRAUNSCHWEIG

Pergamon Press Ltd., Headington Hill Hall, Oxford
4 & 5 Fitzroy Square, London W.1
Pergamon Press (Scotland) Ltd., 2 & 3 Teviot Place, Edinburgh 1
Pergamon Press Inc., Maxwell House, Fairview Park, Elmsford, New York 10523
Pergamon Press (Aust.) Pty. Ltd., 19a Boundary Street, Rushcutters Bay,
N.S.W. 2011, Australia
Pergamon of Canada Ltd., 207 Queen's Quay West, Toronto 1
Pergamon Press S.A.R.L., 24 rue des Ecoles, Paris 5^e
Vieweg & Sohn GmbH, Burgplatz 1, Braunschweig

Copyright © 1969
Pergamon Press Inc.

All Rights Reserved. No part of this publication may be reproduced, stored in a retrieval system, or transmitted, in any form or by any means, electronic, mechanical, photocopying, recording or otherwise without the prior permission of Pergamon Press Limited.

First edition 1969

Library of Congress Catalog Card No. 79–80843

PRINTED IN GREAT BRITAIN BY
ADLARD AND SON LTD.
BARTHOLOMEW PRESS, DORKING, ENGLAND

08 013895 0

CONTENTS

Preface   ix

KEYNOTE ADDRESS

Recent Progress in the Physics of Selenium and Tellurium   3
J. STUKE

BAND STRUCTURE

Band Structure Calculations on Selenium and Tellurium   23
O. MADELUNG and J. TREUSCH

Magnetoabsorption in Tellurium   31
C. RIGAUX, G. DRILHON and Y. ALPERT

Shubnikov–de Haas Effect in Tellurium   47
C. GUTHMANN and J. M. THUILLIER

Interband Transition of Holes in Tellurium   53
D. HARDY and C. RIGAUX

Trapping Levels in Hexagonal Selenium   59
H. P. D. LANYON and R. M. KRAMBECK

Pseudopotential Band Structure for Selenium   69
R. SANDROCK

Optical Properties and Energy Band Structure of Trigonal Selenium   75
W. HENRION

CRYSTAL GROWTH AND CHARACTERIZATION

Impurities in Selenium—Detection and Influence on Physical Properties   87
H. GOBRECHT

Single-crystal Growth of Hexagonal Selenium from Impurity Doped Melts   103
R. C. KEEZER

Etch Pit Studies on Single Crystals of Hexagonal Selenium Grown from the Melt at High Pressures   115
J. D. HARRISON and D. E. HARRISON

The Growth of Selenium Single-crystal Films C. H. GRIFFITHS and H. SANG	135
Solubility and Growth of Trigonal Selenium from Aqueous Sulfide Solutions E. D. KOLB	155
The Morphology and Growth of Trigonal Selenium Crystals C. H. GRIFFITHS and BRIAN FITTON	163
The Growth of α- and β-Red Monoclinic Selenium Crystals and an Investigation of Some of Their Physical Properties G. B. ABDULLAYEV, Y. G. ASADOV and K. P. MAMEDOV	179
Preparation and Identification of Selenium α-Monoclinic Crystals Grown from Selenium-saturated CS_2 S. IIZIMA, J. TAYNAI and M-A. NICOLET	199
Crystallization and Viscosity of Vitreous Selenium H. P. D. LANYON	205
Growth, Perfection and Damageability of Tellurium Single Crystals E. D. KOLB and R. A. LAUDISE	213
Coordination and Thermal Motion in Crystalline Selenium and Tellurium P. UNGER and P. CHERIN	223

OPTICAL PROPERTIES

Infrared-active Lattice Vibrations in Amorphous Selenium A. TAUSEND	233
Optical Properties of Amorphous and Liquid Selenium A. VAŠKO	241
The Structure of Amorphous Selenium from Infrared Measurements G. LUCOVSKY	255
The Raman Spectrum of Trigonal, α-Monoclinic and Amorphous Selenium A. MOORADIAN and G. B. WRIGHT	269
The Phonon Spectra of Trigonal Selenium and Tellurium R. GEICK and U. SCHRÖDER	277
Photoluminescence of Selenium and Selenium–Tellurium Mixed Crystals H. J. QUEISSER	289
Light Beam Modulation in Trigonal Selenium J. E. ADAMS and W. HAAS	293

Investigations on Crystalline Tellurium and Solid Amorphous and
Liquid Selenium with Inelastic Neutron Scattering 299
A. Axmann, W. Gissler and T. Springer

The Contribution of the Lattice Vibrations to the Optical Constants of
Tellurium 309
R. Geick, P. Grosse and W. Richter

ELECTRICAL PROPERTIES

Investigation of the Charge-transfer State in Selenium by the Electronic
Paramagnetic Resonance Method 321
G. B. Abdullayev, N. I. Ibragimov and Sh. V. Mamedov

The Conductivity Mechanism in Monocrystalline Selenium 335
T. Salo, T. Stubb and E. Suosara

Electrical Conductivity of Selenium and Selenium-containing Glasses
at Temperatures up to 1000°C 345
R. W. Haisty and H. Krebs

Electrical Behavior of the Contact Between Cadmium and a Single
Crystal Selenium Film 349
C. H. Champness, C. H. Griffiths and H. Sang

Magnetophonon Resonance in Tellurium 371
D. V. Mashovets and S. S. Shalyt

Author Index 375
Subject Index 377

PREFACE

The Physics of Selenium and Tellurium is the Proceedings of an international symposium on this subject which was held October 12th and 13th, 1967, in Montreal, Canada, under the sponsorship of the Selenium–Tellurium Development Association. The Symposium brought together many of the scientists who have made notable contributions to the Physics of Selenium and Tellurium. The Symposium was a natural outgrowth of the Selenium Physics Symposium which was held in London, England, in June 1964 under the auspices of the European Selenium–Tellurium Committee.

Clearly at a two-day symposium it was not possible to cover all aspects of the subject. Emphasis was placed on the following areas: band structure, crystal growth and characterization, optical properties and electrical properties.

This publication is particularly significant since it gathers together in one volume the important advances which have been made in recent years in the physics of selenium and tellurium. Of particular importance is the work which is reported on the growth and properties of single-crystal selenium, either as bulk crystals or as thin films. Only in recent years have single crystals of selenium become available and some of their properties determined. The preparation of single crystals of tellurium is also discussed, together with a consideration of band structure and transport properties.

This publication can be considered as a companion volume to the Proceedings of the 1964 Symposium published by Pergamon Press under the title *Recent Advances in Selenium Physics*. It is hoped that the present publication will serve as an important stimulus to further research, as was the case with the earlier volume.

The Selenium–Tellurium Development Association is to be commended on sponsoring the Symposium and facilitating the publication of the Proceedings. Support from the National Research Council, Ottawa, Canada, is gratefully acknowledged. A special acknowledgement is due Professor C. H. Champness, McGill University, Montreal, Canada, Dr. H. I. Fusfeld, Kennecott Copper Corporation, New York, and Dr. W. W. Harvey, Ledgemont Laboratory, Kennecott Copper Corporation, Lexington, Mass., for assisting so ably in the editing of the manuscripts.

Noranda Research Centre,
Pointe Claire, Quebec, Canada

W. CHARLES COOPER
Editor and Symposium Chairman

KEYNOTE ADDRESS

RECENT PROGRESS IN THE PHYSICS OF SELENIUM AND TELLURIUM

J. STUKE

Physikalisches Institut (II) der Universität Marburg, Marburg/Lahn, Germany

IN THIS initial symposium paper the recent progress in the physics of selenium and tellurium is reviewed. The paper deals mainly with the results which have been obtained since the Selenium Physics Symposium held in London in 1964.[1] Only a rather brief review can be given because it is not possible to mention all of the interesting results obtained during this period.

Trigonal (sometimes called hexagonal) selenium and tellurium with their chain lattices and strong anisotropy have a number of peculiar and interesting physical properties. The recent results on the different physical properties are discussed for selenium and tellurium together to facilitate a comparison of both elements which is of great value for the understanding of these semiconductors.

Considerable progress has been made in understanding the mechanical and thermal properties of trigonal selenium and tellurium and their lattice dynamics, particularly for selenium. In Fig. 1 the elastic constants of both elements have been compiled. The values for tellurium are those obtained by Malgrange,

Elastic stiffness constants	Se	Te	$\dfrac{C_{\mu\nu}(\text{Se})}{C_{\mu\nu}(\text{Te})}$
	10^{11} dynes/cm^2		
C_{11}	1.87	3.27	0.57
C_{12}	0.71	1.37*	0.52
C_{13}	2.62	2.50	1.05
$|C_{14}|$	0.62	1.23	0.50
C_{33}	7.41	7.22	1.03
C_{44}	1.49	3.14	0.48
Anisotropy	Se	Te	Ratio
C_{33}/C_{11}	4.0	2.2	1.82
C_{13}/C_{12}	3.5	1.8	1.94
$\kappa_\parallel/\kappa_\perp$	3.7	2.0	1.85

*Ref. (3).

FIG. 1. Comparison of the elastic stiffness constants of trigonal selenium[4] and tellurium.[2,3] In the lower part the anisotropy of elastic constants and of thermal conductivity[6] are compared.

3

Quentin and Thuillier[2] except the value of C_{12}, where a new result of Vedam[3] has been used. The constants for selenium have been measured by Mort,[4] who succeeded in overcoming the experimental difficulties which arise from the strong attenuation of the lattice modes perpendicular to the c-axis. The knowledge of the values for this direction now allows a comparison of both elements concerning their mechanical properties. In the last column the ratio of the constants of selenium to those of tellurium are plotted. The ratio is for the constants C_{13} and C_{33}, which are due mainly to the strong covalent bond within the chains, slightly higher than 1. This value might be a little greater because the constants C_{33} and C_{44} found for selenium by Vedam, Miller and Ray[5] are about 5 per cent higher than those of Mort. The bonding strength within the chains obviously is not much greater for selenium than for tellurium. The constants C_{11}, C_{12}, C_{14} and C_{44}, on the other hand, where the weaker bonding between the chains is characteristic, are for selenium only about half of the values found for tellurium. This result indicates that the bonding strength between the chains is for tellurium roughly twice as great as that for selenium.

Another interesting problem is the anisotropy of the bonding. This is revealed by the ratio of different constants for the same element. The anisotropy of the longitudinal constants C_{33}/C_{11} has for selenium the value of 4, for tellurium 2.2. The anisotropies for the shear constants C_{13}/C_{12} are 3.5 and 1.8 respectively. Therefore the bonding anisotropy of selenium is also roughly twice as great as that of tellurium. Its value is determined mainly by the different bonding between the chains, because the bonding strength within the chains is not very different for both elements. The absolute values of the anisotropy suggest that the bonding between the chains is not only of the van der Waals type. The relatively strong bonding particularly for tellurium probably has to be explained by a weak overlap of wave functions.

The anisotropy is likewise revealed by the lattice dynamics and the thermal properties, the thermal conductivity for instance. In the high-temperature range, where phonon–phonon interaction predominates, the thermal conductivity of selenium crystals parallel to the c-axis is greater by 3.7 than that perpendicular to the chains. For tellurium this anisotropy amounts to 2.0.[6] It is interesting to note that these values are similar to the anisotropy of the elastic constants. The similarity is also exhibited by the ratios of the anisotropies which are approximately equal. In Fig. 2 the anisotropy of the thermal conductivity κ is to be seen in more detail. Here κ is plotted versus the absolute temperature, both logarithmically. Below these curves the ratio of the conductivity parallel to the c-axis to that perpendicular to the chains is shown.[7] For tellurium this ratio has in the phonon–phonon interaction region a value of 2 which does not depend on temperature. The increase with decreasing temperature at 5°K is mainly due to the different location of the maxima. For selenium the anisotropy is also approximately independent of temperature

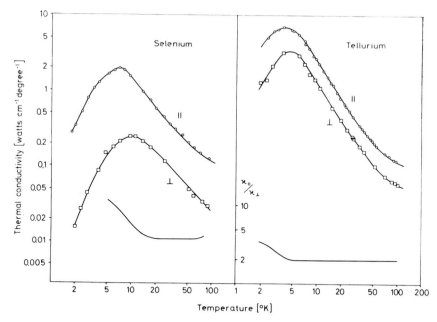

Fig. 2. Temperature dependence of thermal conductivity of trigonal selenium and tellurium.[6, 7]

with a value of 3.7. But the increase starts already at about 15°K because the maxima are located at higher temperatures. This difference can be explained by the smaller size and the minor quality of the selenium crystals compared with tellurium. At low temperatures, where the temperature dependence of the thermal conductivity is determined by that of the specific heat, κ is for selenium approximately proportional to T^3. Since this is the behavior of a normal three-dimensional lattice, at low temperature also the weak bonding between the chains contributes appreciably to the specific heat. According to measurements of Desorbo[8, 9] made some years ago, at higher temperatures the specific heat is proportional only to the temperature T. In this temperature region trigonal selenium and tellurium can be regarded as one-dimensional lattices because the interaction between the chains can be neglected. These two cases are described by two different Debye temperatures, which for selenium amount to $\theta_1 = 370°K$ for the one-dimensional case and $\theta_3 = 75°K$ for the three-dimensional. For tellurium the corresponding values are $\theta_1 = 245°K$ and $\theta_3 = 95°K$.

The results of thermal conductivity and of specific heat give only an integral information about the lattice dynamics. A much more detailed knowledge has been obtained by optical investigations and also by neutron scattering measurements. Particularly significant are the optical measurements of Lucovsky *et al.*[10, 11] The most important results of these investigations and the data obtained by Geick *et al.*[12] and Grosse *et al.*[13] have been compiled in Fig. 3. Here the phonon energies of trigonal selenium and tellurium are shown. They are for selenium evaluated from measurements of the absorption and reflection spectra in the infrared and of the Raman-effect. Approximately the same energies are obtained by measurements of neutron scattering on polycrystalline selenium made by Axmann and Gissler.[14] The phonon energies plotted at the top are not found in the first order optical spectra but by neutron scattering and therefore are acoustical phonons from the edge of the Brillouin zone. The

Acoustical Phonons at the Edge of the Brill. Zone	Se meV	Te meV	Ratio
	5.1	5.3	0.97
	7.8	6.6	1.19

Optical Phonons near $\vec{q} = 0$					
Type of Oscillation	Symmetry Character	Activity Polarisation	Se meV	Te meV	Ratio
	A_2	IR($\parallel$)	12.8	11.0	1.16
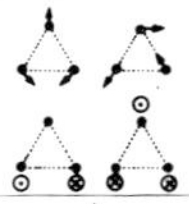	E	IR($\perp$),Ra	17.6	11.7	1.50
		IR($\perp$),Ra	29.0	17.8	1.63
	A_1	Ra	29.5		

FIG. 3. Phonon energies of trigonal selenium and tellurium. Acoustical phonons taken from (14), optical phonons from (10–12) and (13).

other four phonons are zone center optical phonons belonging to different types of oscillations which are indicated in the first column. The first of them is a rotational oscillation of the rigid chains with symmetry A_2 and energy 12.8 meV. It is strongly infrared active for light polarized parallel to the *c*-axis. The other simple type of oscillation with symmetry A_1 has the highest energy of 29.5 meV since it is due to stretching of the covalent bonds within the chains. Because of its symmetry this phonon is only Raman active or it is found in the luminescence spectrum by phonon replica.[15] The other two optical phonons correspond to the more complicated oscillations with symmetry *E*. They are superpositions of these two types of oscillations which are each doubly degenerate.[12] The phonon with energy 17.6 meV is strongly infrared active for light polarized perpendicular to the *c*-axis and also Raman active. Here shearing of the covalent bond predominates. The third one with energy 29.0 meV is essentially created by stretching of the covalent bond. It is weakly infrared active and also Raman active.

The corresponding values for tellurium are also evaluated from recent infrared absorption and reflectivity measurements on single crystals and from neutron scattering on polycrystalline tellurium. With the exception of the first acoustical phonon the energies of tellurium are smaller than those of selenium. The ratios of the selenium–tellurium energies deviate from the value 1.27

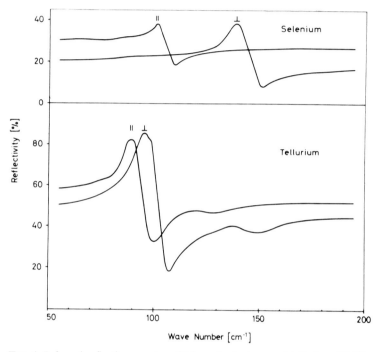

FIG. 4. Infrared reflection spectra of trigonal selenium and tellurium.[10–13]

which one obtains on the basis of the mass difference of both elements only. It is the square root of the mass ratio. For the acoustical phonons and the first optical phonon for which the weak bonding between the chains predominates, the ratio is lower than 1.27 because this bonding increases in its strength from selenium to tellurium. The ratio is higher for those phonons which are determined mainly by the covalent bond within the chains.

For tellurium the two optical phonons with lowest energy are optically even more active than for selenium. This high activity of the trigonal lattice is due to its lack of a center of symmetry and to the high polarizability of the chains because of atomic deformations. Therefore, these oscillations create a strong dipole moment. The effective charge and the oscillator strength have such a high value, particularly for tellurium, that strong reststrahlenbands, as shown in Fig. 4, are found in the reflection spectra of both elements.[10, 11, 13] This is an interesting result because such spectra so far have only been obtained for materials of ionic or partly ionic character. The reflectivity due to the polarization reaches for tellurium values greater than 80 per cent indicating the very high polarizability for this element. For selenium the values are appreciably smaller.

The polarization of the chains exhibited in the optical spectrum is also the reason for the strong piezoelectric effect of both elements. For selenium this was detected several years ago by Gobrecht, Hamisch and Tausend.[16] For tellurium piezoelectricity was found in 1964 by Quentin and Thuillier.[17] These properties are the reason for a number of other interesting results which can only be mentioned briefly. The interaction between these lattice modes and free carriers is of special interest. It has been studied by acoustoelectric phenomena, by current saturation and current instability[18-21] and by acoustical amplification.[22] The combination of high piezoelectric constants and high values of the refractive index leads to a large nonlinearity of the optical constants in the infrared which has been used by Patel[23] for tellurium to create very strong second harmonics. For selenium the electro-optic effect was found by Teich and Kaplan.[24]

For all of these problems the whole phonon spectrum of both elements is of great interest. The first calculations were made for tellurium by Hulin[25] some years ago. His results, which are restricted to the c-direction, are in reasonable qualitative agreement with experiments, but the quantitative values of the calculated frequencies deviate from the recent experimental results indicating that the force constant model used by Hulin is still too simple. Recently Geick et al.[12] made similar calculations for selenium which extended also to the two directions perpendicular to the c-axis. The result for one of these directions is shown in Fig. 5. The details of this rather complicated spectrum are discussed by Geick in his paper in the Symposium Proceedings.[26] Here it shall only be shown that the experimentally found energy values of the fundamental optical lattice modes and also of the acoustical phonons fit into this model with reasonable accuracy. The full circles are the two optically active phonons in the center of the Brillouin zone. The acoustical phonon of 7.8 meV obviously is located at Z whereas the lowest acoustical phonon seems to fit to the high density of states around A and Q. Certainly this model still needs further improvement particularly by inclusion of the Coulomb interaction. But nevertheless it is progress to have now a complete phonon spectrum for selenium. Similar results for tellurium have been obtained very recently.[26] Because of the higher polarizability of tellurium it is even more important than for selenium to take the Coulombic forces into account.

Before proceeding to the recent electrical and other optical results for trigonal selenium and tellurium, mention must be made of the progress in the growth of selenium crystals. At the London conference Harrison[1, 27] presented for the first time sizable selenium crystals grown under hydrostatic pressure at 5000 atm. These crystals of about 10 cm length and 1 cm diameter had dimensions which were unknown for selenium up to that time. Etch pit investigations on these high-pressure crystals are discussed in this volume.[28] During the last two years considerable further progress has been made,

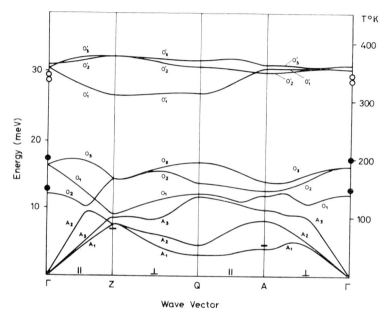

FIG. 5. Phonon spectrum of trigonal selenium.[12]

particularly by Keezer[29-31] and his co-workers. Keezer succeeded in growing relatively large crystals of a high quality by several methods. Everyone who has tried to grow such selenium crystals knows how difficult this is. Some of the important electrical and optical results would not have been obtained without these crystals. An interesting technique has been developed by Lemercier and Thuillier,[32] who used tellurium single crystals as seed crystals for selenium. By epitaxial growth from the melt selenium single crystals of a reasonable size and of good quality have been grown. Further progress in this field has been made by Griffiths and Sang,[33] who obtained very thin single crystalline layers by epitaxial growth from the vapour phase. These thin crystals are of great interest for optical measurements in the intrinsic absorption region. With these layers the first single crystal selenium rectifier has been made by Champness et al.,[34] a result which is certainly of importance for a better understanding of the technical polycrystalline selenium rectifier.

In consideration of recent electrical and optical results, it is noteworthy that progress has been made in the case of selenium whose transport mechanism has been problematical for a long time, because an exponential temperature dependence of the Hall mobility is found for this element. Recently it has been shown by two different methods that the lattice mobility of the holes in trigonal selenium has a temperature dependence like that of tellurium. This is indicated by measurements of the magnetoconductivity by Mell and Stuke[35] and by the study of the acoustoelectrical effect by Mort.[18] Both methods indicate that the hole mobility μ_p decreases with increasing temperature approximately proportional to $T^{-3/2}$. The absolute value of μ_p is rather low, namely about 20 cm^2/Vsec at room temperature parallel to the c-axis. For the direction perpendicular to the chains the mobility is even lower by a factor of about 3.5.[18, 35] The transport mechanism of tellurium has been well known for several years, particularly by the extensive work of Shalyt and his group.[36-39] The hole mobility at room temperature amounts to approximately 2000 cm^2/Vsec. It increases with decreasing temperature for high quality crystals to values as high as 10^5 cm^2/Vsec. In spite of this two orders of magnitude difference compared with selenium, the conductivity mechanism of both elements is similar. The mobility of the holes is determined by phonon scattering. Besides scattering at acoustical phonons, the optical phonons seem to influence the mobility of tellurium. For selenium this interaction is probably smaller.[10, 11] The relatively low mobility of this element obviously is not simply correlated to the strong polarizability of the chains because the polarization is appreciably greater for tellurium which has a much higher mobility. Therefore the low hole mobility of trigonal selenium seems to be due mainly to a high value of the effective mass.

Recent experimental results have provided new information about the band structure of trigonal selenium and tellurium. Since selenium and tellurium are p-type in the extrinsic conduction region, the first important problem is the structure of the valence band. For both elements the hole mobility parallel to the c-axis $\mu_\parallel$ is greater than $\mu_\perp$. For selenium the ratio $\mu_\parallel/\mu_\perp$ amounts to about 3.5[18, 35] and for tellurium to 1.25.[36-39] It is determined by two components, namely the ratio of the relaxation times $\tau_\parallel/\tau_\perp$ divided by the ratio of the effective masses $m_\parallel/m_\perp$. For selenium the effective mass in the direction of the chains seems to be appreciably smaller than the mass perpendicular to the c-axis. This can be concluded from the anisotropy of the Franz–Keldysh effect.[40] For tellurium just the reverse character of the mass ratio has been found, at least at low temperatures. At 4°K effective mass in the chain direction is greater by a factor 2.4 than the mass perpendicular to the c-axis. This has been shown by recent measurements of cyclotron resonance[41, 42] and of Shubnikov–de Haas-effect.[43] Taking into account the mobility ratio this result indicates that the relaxation time is strongly anisotropic, the ratio $\tau_\parallel/\tau_\perp$ having a value of 3. This anisotropy of the relaxation time and the mass ratio of the valence band are very interesting results. A more detailed discussion of the mass anisotropy will be given a little later when some experimental and theoretical band structure features are compared.

Another important problem is the bandgap of trigonal selenium and tellurium, its size and its properties. For tellurium the value of the gap is now known with high accuracy by optical measurements. The results on magneto-absorption obtained by Rigaux and Serrero[44–46] and by Winzer and Grosse[47] yield a value of 0.334 eV for both polarizations. It is a direct gap with allowed

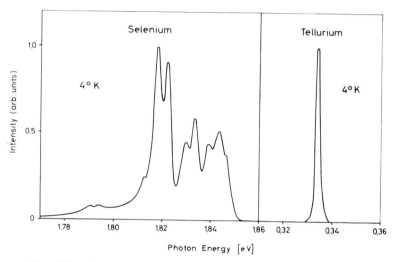

FIG. 6. Luminescence spectra of trigonal selenium[49] and tellurium.[48]

transitions for light polarized perpendicular to the c-axis and forbidden transitions for parallel polarization. For both polarizations excitons are found at the absorption edge. The direct character of the gap is also exhibited by luminescence measurements. In Fig. 6 these results are shown on the right-hand side. Benoit à la Guillaume and Debever[48] found a relatively sharp emission at 0.334 eV which is polarized perpendicular to the c-axis. At higher injection levels stimulated emission has been obtained. This character of the luminescence indicates conclusively that for tellurium the emission is due to direct band-to-band recombination.

Recently for selenium luminescence has also been found at 20°K.[15] On the left-hand side new results obtained at 4°K by Queisser[49] are plotted. It is easily seen that the luminescence spectrum of selenium has a quite different character. Obviously transitions between local levels or excitons are involved. Therefore no direct conclusion on the bandgap of trigonal selenium can be

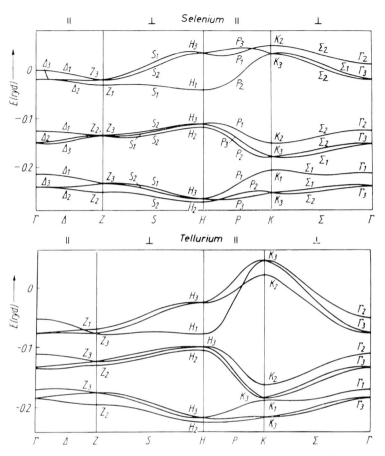

FIG. 7. Band structure of trigonal selenium and tellurium.[55]

drawn from these investigations, but its value can be derived from reflectivity measurements. Similarly to tellurium an exciton is found for light polarized perpendicular to the c-axis at 1.95 eV.[50-52] The bonding energy of this exciton is estimated to be approximately 0.05 eV. Therefore the value of the bandgap is 2.00 eV for perpendicular polarization. This gap is also a direct one. For the polarization parallel to the c-axis the situation is not yet quite clear. Here no exciton has been found in the reflection spectrum. The structure of the absorption edge seems to indicate indirect transitions,[53] but further evidence is necessary to make this result conclusive.

Madelung has reviewed the band structure calculations and their results.[54] Figure 7 shows as an example the band structure obtained by Treusch and Sandrock[55] for selenium and tellurium. A direct gap is found for both elements at the edge of the Brillouin zone at or near the point H. This feature is consistent with most of the optical results found for both elements, particularly with the existence of an intervalence band absorption (p-band) for tellurium.[56, 57] The p-band with its polarization parallel to the c-axis is consistent with theory if the maximum of the valence band is located at the point H. This conclusion can be attained by symmetry arguments and is important because it does not depend on numerical values which may be uncertain.

The perpendicular polarization of the allowed direct transitions across the band gap together with the parallel polarization of the p-band strongly suggest that the gap of tellurium is located at the point H. The calculated mass ratio, on the other hand, has at H the wrong character. This feature can be easily seen in Fig. 7 because the curvature of the valence band in the c-direction is appreciably greater than that perpendicular to the c-axis. Other band-structure calculations gave qualitatively the same result.[58, 59] Therefore at the point H the effective hole mass parallel to the c-axis theoretically is smaller than that perpendicular to it. This leads to a mass ratio which is just the reverse of the experimentally found character. The latter type of ratio on the other hand occurs in the center of the Brillouin zone. Here the curvature in the c-direction is smaller than that perpendicular to it. Therefore the valence band in Γ has the experimentally found character of the mass ratio for tellurium, but a p-band cannot be explained there because of symmetry. Since after recent measurements by Grosse and Selders[60] the p-band is experimentally also found at $4°K$ with a normal height, this dilemma cannot be solved by the assumption that with decreasing temperature the gap is shifted from H to Γ. The gap of tellurium obviously stays in H, but the theoretically found mass ratio $m_\parallel/m_\perp$ is wrong. More accurate calculations near the gap at H which include also the influence of the temperature on the lattice constants should be made to solve this fundamental problem. Recent investigations of Grosse and Lutz[61] seem to indicate that the mass perpendicular to the c-axis decreases appreciably with decreasing temperature, whereas $m_\parallel$ is approximately independent of temperature. A similar result has already been found by Caldwell and Fan.[56]

A peculiar feature of the band structure of selenium and tellurium is an energy gap which exists between the two groups of valence bands. This gap leads to a minimum in the optical spectra at high energies, the reflection spectra, for instance, because the combined density of states for transitions from both triplets of valence bands to the conduction bands is low in the region of this gap. Figure 8 shows this behavior for selenium and tellurium. The curves for tellurium above 6 eV are embodied in the yet unpublished results of Cardona.[62] The data below 5 eV are taken from measurements of Stuke and Keller.[63] The curves for selenium are the results of Mohler, Stuke and Zimmerer.[64] For light polarized perpendicular to the c-axis the minimum is easily seen for both elements. Selenium has also a minimum for the parallel polarization. It is followed by a strong maximum which is relatively broad and which is only observed for parallel polarization. For tellurium a similar maximum does not seem to exist at first sight since only a small hump in the curve for parallel polarization is observed. But the curves of the ratio of the reflectivity values for both polarizations reveal that the behavior of tellurium is quite similar. In both curves pronounced maxima occur at or near the

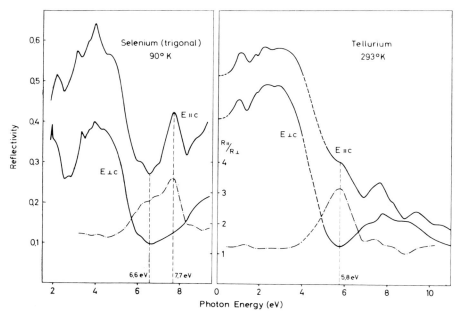

FIG. 8. Reflection spectra of trigonal selenium[64] and tellurium.[62, 63]

reflectivity minima. Therefore for both elements the main optical anisotropy is located at high energies. This very strong anisotropy obviously is correlated to transitions from the lower triplet of the valence bands to the conduction bands. The strength of this peak and its width indicate that the relevant transitions are not restricted to critical points in the Brillouin zone but occur along axes within the zone. The parallel polarization is consistent with transitions along the Δ-axis. This is an interesting feature which has not been found to such a degree for other semiconductors.

The properties mentioned so far are mainly those of the intrinsic material. Now some new results on the influence of impurities and their determination will be discussed. Several papers presented at this Symposium deal with these problems. The paper of Gobrecht[65] shows that the content of some important impurities in selenium, particularly of chlorine, can be determined by radiochemical methods with a high accuracy. Impurity absorption bands in the infrared have been used by Tausend[66] and by Vasko[60] for a quantitative determination of the oxygen content. A more detailed knowledge of the character of some impurities within the selenium lattice has been obtained by measurements of paramagnetic resonance made by Abdullayev and co-workers[68, 69] and by Sampath[70] and Chen.[71] The influence of these impurities on the electrical properties of trigonal selenium crystals is still problematical. Oxygen seems to be of particular interest, but lattice defects also have a great influence on electrical properties. Dangling bonds due to vacancies or free chain ends act as acceptors in selenium. Their density normally amounts to about 10^{14} to 10^{15} cm^{-3} and can be strongly enhanced by plastic deformation.[72] Also for tellurium, acceptors of this type obviously are of importance. The acceptor concentration of 10^{13} cm^{-3} which seems to be a lower limit may be due to such lattice defects which are created by thermal treatment. Recent quenching investigations made by Hörstel and Kretschmar[73] indicate that at high temperature the density of thermal acceptors is strongly enhanced. It increases exponentially with temperature with an activation energy of 0.8 eV. It is therefore very possible that at high temperature the p-type conduction due to these acceptors predominates and that this is the reason for the second sign reversal of the Hall effect and thermoelectric power in the intrinsic region.[74] The same explanation for this interesting problem was proposed some years ago by Fritsche[75] and by Tanuma.[76]

Some new results obtained for the other modifications of selenium and tellurium will be reported on briefly. The first one refers to metallic selenium and tellurium. These modifications are stable only under very high pressure. For tellurium 45,000 atm are needed and for selenium even 130,000 atm. Most of the properties of these metallic phases are not yet known because the very high pressure implies great experimental difficulties. Recently Wittig[77] in Karlsruhe succeeded in solving these problems even at 4°K, and he thereby was able to detect super conductivity of metallic selenium at 130,000 atm. His results are shown in Fig. 9. A pronounced decay of the resistance occurs at the critical temperature of about 6.7°K. For tellurium superconductivity of the metallic phase was found some years ago by Matthias and Olsen[78] at 56,000 atm. Here the critical temperature is 3.3°K.

Other modifications of selenium are the monoclinic forms which have a ring structure and are therefore similar to sulphur. Abdullayev[79] has investigated the growth of α- and β-monoclinic selenium crystals and some of their physical properties. Lucovsky et al.[10, 11] have determined the fundamental lattice modes of α-monoclinic selenium from absorption and Raman spectra. At low energies relatively sharp absorption bands are found which are of importance not only for the lattice dynamics of monoclinic selenium but also for the investigations on the structure of amorphous selenium.[80] Namely, these absorption bands are also found in the absorption spectrum of the amorphous phase. Lucovsky *et al.* thereby have been able to show conclusively, that amorphous selenium has a high content of rings like monoclinic selenium. This result is of great importance for the electrical and photoelectrical properties of this material. Finally a result for amorphous selenium discussed at this Symposium is the change of the neutron scattering spectrum, if one goes from the crystalline to the amorphous or to the liquid state. This can be seen in Fig. 10 which shows recent results of Axmann, Gissler and Springer.[81] Whereas the maxima 3, 4 and 5 due to the optical phonons are only slightly

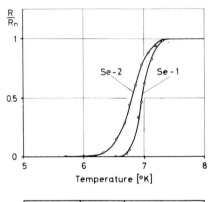

Element	T_C °K	Pressure atm
Se	6.7	130000
Te	3.3	56000

FIG. 9. Superconductivity of metallic selenium[77] and tellurium.[78]

changed in height and energy, the two acoustical phonon peaks 1 and 2 disappear. This result indicates that the short range order is approximately the same in the crystalline and in the amorphous state. The long-range order, on the other hand, which is essential for the acoustical phonons is strongly disturbed. Similar investigations have been made by Regel and co-workers[82] in Leningrad. Infrared absorption and neutron scattering obviously are very useful for the study of the amorphous state.

In conclusion it has to be emphasized that a number of other interesting properties of selenium and tellurium have not been mentioned, for instance photoconductivity, lattice-diffusion or frequency dependence of conductivity. The latter effect is discussed by Stubb[83] for trigonal selenium. It is hoped that the remarkable progress in the physics of selenium and tellurium has been made evident. For tellurium the gradient of this progress is a little smaller than

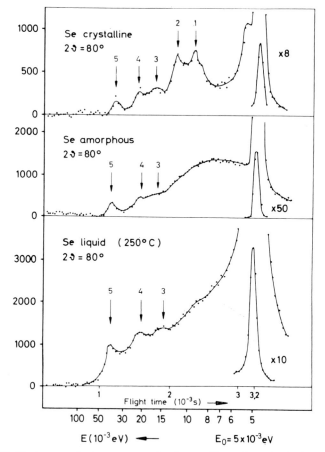

FIG. 10. Neutron scattering spectra of polycrystalline trigonal selenium and of amorphous and liquid selenium.[81]

for selenium because tellurium belongs already to the family of well acknowledged semiconductors. For selenium, however, this international conference on the physics of selenium and tellurium will very probably open a new phase in the understanding of this important and interesting material.

REFERENCES

1. *Recent Advances in Selenium Physics*, Pergamon, London, 1965.
2. MALGRANGE, J. L., QUENTIN, G. and THUILLIER, J. M., *Phys. Status Solidi* **4**, 139 (1964).
3. VEDAM, K., private communication.
4. MORT, J., *J. Appl. Phys.* **38**, 3414 (1967).
5. VEDAM, K., MILLER, D. L. and RAY, R., *J. Appl. Phys.* **37**, 3432 (1966).
6. ADAMS, A. R., BAUMANN, F. and STUKE, J., *Phys. Status Solidi* **23**, K99 (1967).

7. ADAMS, A. R., BAUMANN, F. and STUKE, J., to be published.
8. DESORBO, W., *J. Chem. Phys.* **21**, 764 (1953).
9. DESORBO, W., *J. Chem. Phys.* **21**, 1144 (1953).
10. LUCOVSKY, G., MOORADIAN, A., TAYLOR, W., WRIGHT, G. B. and KEEZER, R. C., *Solid State Comm.* **5**, 113 (1967).
11. LUCOVSKY, G., KEEZER, R. C. and BURSTEIN, E., *Solid State Comm.* **5**, 439 (1967).
12. GEICK, R., SCHRÖDER, M. and STUKE, J., *Phys. Status Solidi* **24**, 99 (1967).
13. GROSSE, P., LUTZ, M. and RICHTER, W., *Solid State Comm.* **5**, 99 (1967).
14. AXMANN, A. and GISSLER, W., *Phys. Status Solidi* **19**, 721 (1967).
15. QUEISSER, H. J. and STUKE, J., *Solid State Comm.* **5**, 75 (1967).
16. GOBRECHT, H., HAMISCH, H. and TAUSEND, A., *Z. Physik* **148**, 209 (1957).
17. QUENTIN, G. and THUILLIER, J. M., *Solid State Comm.* **2**, 115 (1964).
18. MORT, J., *Phys. Rev. Letters* **18**, 540 (1967).
19. QUENTIN, G. and THUILLIER, J. M., *Proc. Int. Conf. on The Physics of Semiconductors*, Dunod, Paris, 1964.
20. QUENTIN, G. and THUILLIER, J. M., *Proc. Int. Conf. on The Physics of Semiconductors*, Kyoto, The Physical Society of Japan, Tokyo, 1966, p. 493.
21. ISHIGURO, T., NITTA, S., HOTTA, A. and TANAKA, T., *J. Appl. Phys. Japan* **4**, 703 (1965).
22. ISHIGURO, T. and TANAKA, T., *Proc. Int. Conf. on The Physics of Semiconductors*, Kyoto, The Physical Society of Japan, Tokyo, 1966, p. 489.
23. PATEL, C. K. N., *Phys. Rev. Letters* **15**, 1027 (1965).
24. TEICH, M. C. and KAPLAN, T., *J. Quantum Electron.* **2**, 702 (1966).
25. HULIN, M., *Ann. Phys. (Paris)* **8**, 648 (1963).
26. GEICK, R. and SCHRÖDER, U., *The Physics of Selenium and Tellurium*, Pergamon, New York, 1969, p. 277.
27. HARRISON, D. E. and TILLER, W. A., *J. Appl. Phys.* **36**, 1680 (1965).
28. HARRISON, J. D. and HARRISON, D. E., *The Physics of Selenium and Tellurium*, Pergamon, New York, 1969, p. 115.
29. KEEZER, R. C. and WOOD, C., *Appl. Phys. Letters* **8**, 139 (1966).
30. KEEZER, R. C. and MOODY, J. W., *Appl. Phys. Letters* **8**, 233 (1966).
31. KEEZER, R. C., WOOD, C. and MOODY, J. W., *Proc. Int. Conf. on Crystal Growth*, Pergamon, New York (1967).
32. LEMERCIER, C. and THUILLIER, J. M., *Mat. Res. Bull.* **1**, 109 (1966).
33. GRIFFITHS, C. H. and SANG, H., *The Physics of Selenium and Tellurium*, Pergamon, New York, 1969, p. 135.
34. CHAMPNESS, C. H., GRIFFITHS, C. H. and SANG, H., *The Physics of Selenium and Tellurium*, Pergamon, New York, 1969, p. 349.
35. MELL, H. and STUKE, J., *Phys. Letters* **20**, 222 (1966); *Phys. Status Solidi*, to be published.
36. PARFEN'EV, R. V., FARBSHTEIN, I. I. and SHALYT, S. S., *Sov. Phys. Sol. State* **2**, 2599 (1961).
37. PARFEN'EV, R. V., POGARSKII, A. M., FARBSHTEIN, I. I. and SHALYT, S. S., *Sov. Phys. Sol. State* **3**, 1820 (1962).
38. PARFEN'EV, R. V., POGARSKII, A. M., FARBSHTEIN, I. I. and SHALYT, S. S., *Sov. Phys. Sol. State* **4**, 2630 (1963).
39. FARBSHTEIN, I. I., POGARSKII, A. M. and SHALYT, S. S., *Sov. Phys. Sol. State* **7**, 1925 (1966).
40. STUKE, J. and WEISER, G., *Phys. Status Solidi* **17**, 343 (1966).
41. MENDUM, J. H. and DEXTER, R. N., *Bull. Am. Phys. Soc.* 632 (1964).
42. PICARD, J. C. and CARTER, D. L., *Proc. Int. Conf. on The Physics of Semiconductors*, Kyoto, The Physical Society of Japan, Tokyo, 1966, p. 202.
43. BRAUN, E. and LANDWEHR, G., *Proc. Int. Conf. on The Physics of Semiconductors*, Kyoto, The Physical Society of Japan, Tokyo, 1966, p. 380.
44. RIGAUX, C. and SERRERO, M., *Solid State Comm.* **3**, 21 (1965).
45. RIGAUX, C. and DRILHON, G., *Proc. Int. Conf. on The Physics of Semiconductors*, Kyoto, The Physical Society of Japan, Tokyo, 1966, p. 193.
46. RIGAUX, C., DRILHON, G. and ALPERT, Y., *The Physics of Selenium and Tellurium*, Pergamon, New York, 1969, p. 31.

47. WINZER, K. and GROSSE, P., *Verhandl. DPG* (6) **2**, 29 (1967).
48. BENOIT À LA GUILLAUME, C. and DEBEVER, J. M., *Solid State Comm.* **3**, 19 (1965).
49. QUEISSER, H. J., *The Physics of Selenium and Tellurium*, Pergamon, New York, 1969, p. 289.
50. HENRION, W., *Phys. Status Solidi* **20**, K145 (1967).
51. TUTIHASI, S. and CHEN, I., *Solid State Comm.* **5**, 255 (1967).
52. TUTIHASI, S. and CHEN, I., *Phys. Rev.* **158**, 623 (1967).
53. ROBERTS, G. G., TUTIHASI, S. and KEEZER, R. C., *Solid State Comm.* **5**, 517 (1967).
54. MADELUNG, O. and TREUSCH, J., *The Physics of Selenium and Tellurium*, Pergamon, New York, 1969, p. 23.
55. TREUSCH, J. and SANDROCK, R., *Phys. Status Solidi* **16**, 487 (1966).
56. CALDWELL, R. S. and FAN, H. Y., *Phys. Rev.* **114**, 664 (1959).
57. GROSSE, P. and SELDERS, M., *Phys. Kondens. Materie* **6**, 126 (1967).
58. JUNGINGER, H. G., to be published.
59. PICARD, M. and HULIN, M., to be published.
60. GROSSE, P. and SELDERS, M., *Phys. Status Solidi*, to be published.
61. GROSSE, P. and LUTZ, M., *Verhandl. DPG* (6) **2**, 30 (1967).
62. CARDONA, M., private communication.
63. STUKE, J. and KELLER, H., *Phys. Status Solidi* **7**, 189 (1964).
64. MOHLER, E., STUKE, J. and ZIMMERER, G., *Phys. Status Solidi* **22**, K49 (1967).
65. GOBRECHT, H., *The Physics of Selenium and Tellurium*, Pergamon, New York, 1969, p. 87.
66. TAUSEND, A., *The Physics of Selenium and Tellurium*, Pergamon, New York, 1969, p. 233.
67. VAŠKO, A., *The Physics of Selenium and Tellurium*, Pergamon, New York, 1969, p. 241.
68. ABDULLAYEV, G. M. B., IBRAGIMOV, N. I., MAMEDOV, SH. V., DZHNVARLY, T. CH. and ALIEV, G. M., *Doklady Akad. Nauk Azerb. SSR.* **20**, 13 (1964).
69. ABDULLAYEV, G. M. B., IBRAGIMOV, N. I. and MAMEDOV, SH. V., *The Physics of Selenium and Tellurium*, Pergamon, New York, 1969, p. 321.
70. SAMPATH, P. I., *J. Chem. Phys.* **45**, 3519 (1966).
71. CHEN, I., *J. Chem. Phys.* **45**, 3536 (1966).
72. STUKE, J., *Phys. Status Solidi* **6**, 441 (1964).
73. HORSTEL, W. and KRETSCHMAR, G., *Phys. Status Solidi*, to be published.
74. LINK, R., *Phys. Status Solidi* **12**, 81 (1965).
75. FRITSCHE, H., *Science* **115**, 571 (1952).
76. TANUMA, S., *Sci. Rep. RITU AG*, 159 (1964).
77. WITTIG, J., *Phys. Rev. Letters* **15**, 159 (1965).
78. MATTHIAS, B. T. and OLSEN, J. L., *Phys. Letters* **13**, 202 (1964).
79. ABDULLAYEV, G. M. B., ASADOV, YU. G. and MAMEDOV, K. P., *The Physics of Selenium and Tellurium*, Pergamon, New York, 1969, p. 179.
80. LUCOVSKY, G., *The Physics of Selenium and Tellurium*, Pergamon, New York, 1969, p. 255.
81. AXMANN, A., GISSLER, W. and SPRINGER, T., *The Physics of Selenium and Tellurium*, Pergamon, New York, 1969, p. 299.
82. REGEL, A. R., OKUNEVA, N. M. and KOTOV, B. A., private communication.
83. SALO, T., STUBB, T. and SUOSARA, E., *The Physics of Selenium and Tellurium*, Pergamon, New York, 1969, p. 335.

DISCUSSION

NUSSBAUM: Do you believe that the double reversal could be due to multiple band structure rather than to defects, as indicated in your paper?

STUKE: Several authors have tried to explain the second sign reversal by band structure, either by valence bands and one conduction band or by one valence band and two conduction bands. However, they have not been able to explain with these models the details of the sign reversal of Hall effect and of thermoelectric power together. Link,[74] in a recent discussion of this problem, came to the conclusion that thermally created acceptors can explain the behaviour of Hall effect and thermoelectric power. It might also be possible that the second sign reversal is due to a strong increase of the effective electron mass with the concentration of electrons.[83]

BECKER: Does boundary scattering in low temperature thermal infrared experiments reflect sample dimensions on internal boundaries?

STUKE: Boundary scattering at low temperatures gives the right order of magnitude of the sample dimension but there seems to be also an influence of internal defects.

RICCIUS: The exciton binding energy in selenium was given as 0.05 eV. Is this a calculated value? What is the exciton binding energy in tellurium?

STUKE: The exciton binding energy in selenium has been estimated by Tutihasi and Chen[51, 52] assuming a reasonable value for the effective mass. The exciton binding energy in tellurium amounts to 0.0012 eV according to the results of Rigaux.[44-46]

LUCOVSKY: The neutron spectroscopy curves indicated that the highest energy optical phonon had an energy of about 40 meV, whereas Raman measurements indicate a phonon energy of 29.5 meV. Could you comment on the discrepancy?

STUKE: New measurements of neutron scattering by Gissler and Axmann[14] indicate that the energy values given by these authors recently were somewhat too high. This might be the reason for the discrepancy.

SANDROCK: The location of the gap in the Brillouin zone might be decided upon experimentally by some differential reflectivity method, e.g. piezoreflectivity with pressure $\perp$ c-axis. Is this possible, or is it being done already?

STUKE: Piezoreflectivity measurements with the pressure $\perp$ c-axis should be possible, but there are great experimental difficulties because in this direction selenium crystals are easily plastically deformed or cleaved. We are just trying to overcome these problems.

CHAMPNESS: You mentioned that imperfections cause the low mobilities in single crystal selenium. Would you like to speculate where these are located?

STUKE: This important problem has not been solved yet because we do not know much about the details of the lattice defects in trigonal selenium. I think that the low mobility is correlated to the network of dislocations and that impurities are mainly located in or near these dislocations. Potential barriers are thereby created which determine the temperature dependence of the conductivity.

BAND STRUCTURE

BAND STRUCTURE CALCULATIONS ON SELENIUM AND TELLURIUM

O. MADELUNG and J. TREUSCH

Institut für Theoretische Physik (II) der Universität Marburg, Marburg/Lahn, Germany

IN THE last few years considerable progress has been made in the understanding of the physical properties of selenium and tellurium. Although experimental results will not be discussed in this paper, two of these results can give some characteristic features of the band structure of the two elements.

At the top of the valence band of tellurium two bands must occur split by an amount of 0.1 eV in order to explain the well-known infrared 11 μ-band. In selenium no such absorption band has been assigned conclusively to a band-to-band transition.

Most optical information (at least for tellurium) leads to the conclusion that the transitions at the absorption edge are direct. Then the maximum of the valence band and the minimum of the conduction band lie at the same point in the Brillouin zone.

Selenium and tellurium crystallize in a trigonal structure. The unit cell of the lattice contains three atoms arranged in the characteristic spiral chain. Considering the "chain structure" of these elements one has to keep in mind that besides the chemical bond along the chains there is a weaker but nevertheless considerable bond between the chains. The ratio of the distances between next neighbors in one chain and between two chains is not larger than 1.49 for selenium and 1.2 for tellurium. Thus for tellurium the anisotropy of the structure is still less than for selenium.

Figure 1 shows the Brillouin zone for this structure. By the experience that the relevant extrema of energy bands of semiconductors lie always on points or lines of high symmetry in the Brillouin zone, all calculations up to the present have been restricted to the points Γ, Z, H, K and the axes Δ, S, P, Σ connecting these points. Detailed group theoretical treatment of the symmetry properties of wave functions and energy bands along these axes has been given, for example, by Asendorf.[1]

In the free selenium and tellurium atom the outer electrons are of the type $4p$, $5p$, respectively. Assuming no mixing of the atomic levels, one has to expect three p-bands (consisting of three sub-bands each, since there are three atoms per unit cell) in the relevant region of the band structure. The two lower ones of these three triplets must be occupied by electrons. This has been shown by Reitz[2] in a paper on the band structure of selenium and tellurium as early as 1957, and has been confirmed in all later calculations (see Fig. 2). Considering the binding to the next neighbors only and assuming that there is no mixture of the p-states with the s- and d-states, Reitz calculated the band structure along the Δ-axis of the Brillouin zone by means of the tight binding method. The smallest energy difference between the valence band and the conduction band along this axis was situated in the point Z. Later work showed that it is necessary to extend the calculations along the four axes shown in Fig. 1.

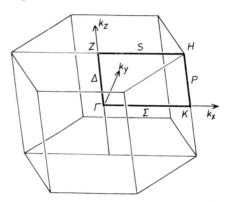

FIG. 1. Brillouin zone of selenium and tellurium.

Reitz's calculations have been improved by Knox and Olechna [3] taking s–p mixing into account. Use was made of non-overlapping atomic potentials and a $\rho^{1/3}$-exchange potential. Considering next neighbors only this calculation was also restricted to the Δ-axis.

The same restriction holds for a tellurium calculation by Beissner,[4] who used the pseudopotential method.

It has been mentioned that calculations of the band structure of tellurium must account for the 0.1 eV splitting of the top of the valence band. None of the non-relativistic calculations of tellurium band structures discussed above gave evidence for two sub-bands of the valence band as close as 0.1 eV satisfying the strong selection rule observed for 11 μ-band. Thus, spin splitting of the uppermost valence band has to be taken into account in order to explain the 11μ-band. This excludes the occurrence of the top of the valence band in tellurium at the points Γ and Z, since the only bands that would satisfy the selection rule are degenerate by time reversal in these points.[5, 6] A valence band maximum on the Δ-axis, where a splitting would be possible, was suggested by Beissner. This suggestion seems unrealistic by several reasons.[7] As far as selenium is concerned, the possibility of the occurrence of the energy gap in Γ, Z or along the Δ-axis cannot be excluded by the above

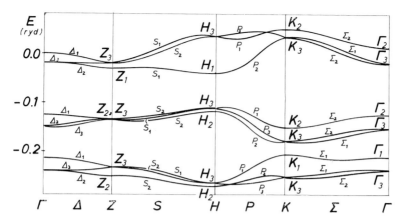

Fig. 2a. Energy band structure according to ref. (6) for selenium.

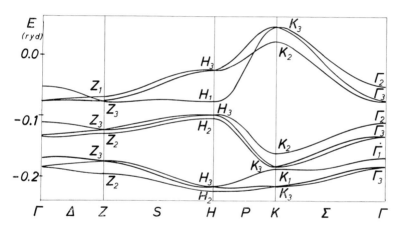

Fig. 2b. Energy band structure according to ref. (6) for tellurium.

arguments. The qualitative similarity of all band structures calculated up to this time for selenium and tellurium, however, makes this possibility very unlikely.

Thus, an extension of the calculations to other axes in the Brillouin zone is necessary. The first step in this direction was made by Hulin,[5] who used the LCAO method (tight-binding), took next and second next neighbors into account and got a band structure for tellurium with an energy gap near the point H at the corner of the Brillouin zone. Here a spin-splitting of the uppermost valence band is possible that leads to two sub-bands with the correct selection rule. A perturbational approach to this spin-orbit splitting gave the right order of magnitude (0.14 eV).

The explanation of an experimental result by a band structure calculated by one of the many possible methods is pleasing. Confidence, however, in such a correspondence can be won only if the band structure can explain more than a single fact or if the relevant details have been shown by different calculations using different methods. Thus it is important to note that most of the recent calculations agree with the result, that the top of the valence band of tellurium lies in the point H. Figure 2 shows as an example the band structure of selenium and tellurium calculated by Treusch and Sandrock[6] by the Green's function method. The energy gap is situated in H for both substances and it is much larger in selenium than in tellurium in agreement with experiment.

Results very similar to those shown in Fig. 2 have been obtained for tellurium by Picard and Hulin [8] with the pseudopotential method. Their tentative results for selenium, however, are only in qualitative agreement with the work of Treusch and Sandrock and with experiment. Further results are given by Sandrock in his Symposium paper.[9] The only theoretical work not consistent with the general trend of the structures shown in Fig. 2 was published by Junginger,[10] who used a modified APW-method for tellurium. The resulting band structure shows the gap at Γ. Besides the valence band extremum in Γ there is another extremum in H of almost equal energy. The improbability of Γ- and H-extrema at the same energy in the full temperature range where the 11 μ-band has been observed, throws some doubt on this result.

The above discussion has concentrated on the explanation of the 11 μ-band by the calculations carried out up to now, for this (besides the direct energy gap) is the most decisive experimental result to be explained by a reasonable band structure. Other facts where experiment and theory can be compared are higher optical transitions and the effective masses (i.e. the curvature of the valence and conduction bands near their extrema).

Effective masses can be obtained from transport phenomena. The only value, however, that can be deduced out of the present band structure calculations on selenium and tellurium with reasonable accuracy is the sign and the order of magnitude of the anisotropy of the effective masses $m_\perp/m_\parallel$. Unfortunately the different experimental methods give no consistent results. Thus no discussion of this problem is given in the present context. Much information about this point is contained in the papers cited above.

Not much better is the situation as to the higher optical transitions. It is possible to assign the energies of critical points in the absorption and reflection spectra to certain band-to-band transitions in band structures like the ones shown in Fig. 2. This is facilitated by use of selection rules for light polarized parallel and perpendicular to the c-axis. The agreement, however, can be only qualitative by experimental and theoretical reasons. Experiments yield only reflection spectra and in most cases there exists not enough information to perform a reliable Kramers–Kronig analysis to transform them to E_2-spectra that can be compared quantitatively with theory. Thus errors of some tenths of an eV in the localization of the critical points are possible. On the other hand, theoretical calculations for the band structures of selenium and tellurium are accurate only to some tenths of an eV. This restriction, of course, does not hold for the spin-orbit splitting, that can be calculated far more exactly. Furthermore, the spin and relativistic effects have not been taken into account quantitatively in the literature discussed above.

An attempt has been made to pay regard to spin-orbit coupling and relativistic effects quantitatively. For this purpose a modification of the Green's function method was used.[11] The details of the calculations will be reported elsewhere.[12] The most important results are:

1. The maximum of the valence band remains at H after inclusion of spin-orbit interaction and the relativistic Darwin- and mass-velocity corrections. The 11 μ-band in tellurium given in the right order of magnitude is almost independent of the potential used (see Fig. 3). In selenium the same splitting is less than 0.05 eV.

2. The Green's function method uses a spherical "muffin-tin" potential, i.e. an atomic potential $V(r)$ inside spheres around the atoms and a constant potential V_o outside the spheres. The value of V_o is often used as an adjustable parameter. As has been shown elsewhere [13] for the cubic II–VI-compounds, fitting of the energy gap by this parameter leads to V_o-values which are very closely the average potential outside the spheres. Moreover, the top of the valence band is then situated approximately at the energy level of the constant potential outside the spheres. Both results are very satisfactory since they give to V_o a physical meaning. The same situation as with the II–VI-compounds is found in the cases of selenium and tellurium. Choosing for V_o a value that meets the requirements discussed above, one obtains for the energy differences between the valence and conduction bands in the Brillouin zone corner H 1.8 eV and 0.33 eV for selenium and tellurium, respectively, thus fitting the experimentally measured energy gap quite well. At this V_o the conduction band minima in Z and Γ decrease to values slightly lower than those in H. The distances between valence and conduction band, however, remain significantly larger in Γ and Z than those in H. In selenium as well as in tellurium a direct (but too small) gap in H can be achieved by another (physically less realistic) choice of V_o. It is felt, however, that the models obtained on the basis of an averaged V_o are more convincing in spite of the discrepancies concerning the conduction bands:

(a) These discrepancies are of an order of magnitude comparable to the inaccuracies unavoidable in any band structure calculation up to now.
(b) The smallest direct gap satisfies experimental data well in the case of both selenium and tellurium.
(c) V_o has a physically reasonable value.

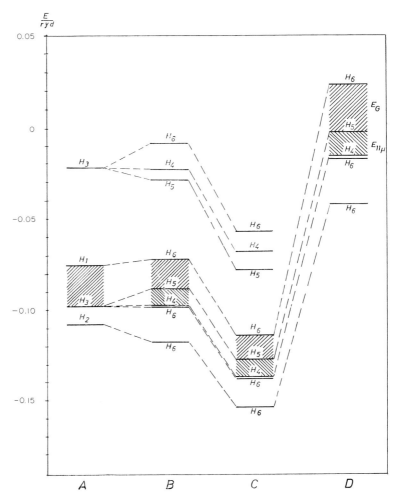

FIG. 3. Spin and relativistic corrections at the Brillouin zone corner H of tellurium. A: non-relativistic case (Fig. 2b). B: spin-orbit coupling included. C: relativistic corrections included. D: valence band shifted to $E=0$ changing V_o.

(d) A slight change of the muffin-tin potential shifts energy bands at different points in the Brillouin zone (Γ, Z, H, K) relative to one another much more strongly than the different energy bands at the same point. Thus the discrepancies mentioned above can probably be overcome by the use of a non-local potential, that depends on l and k. Since in selenium and tellurium the upper valence bands and the lower conduction bands are predominantly p-type, it is felt that the l-dependence of the potential should be negligible. From these considerations it is concluded that in spite of doubtful results in the relative positions of the energy band extrema at different points in k-space, the energy differences at a given point can be trusted. Therefore, as a first order perturbation, a small k-dependent shift of the inter-space potential V_o was introduced, the maximum shift being about 5 per cent of the average value. The resulting band structure for selenium is shown in Fig. 4. It explains the measurements of Tutihasi and Chen[14] more quantitatively than did the band structure of Treusch and Sandrock,[6] but there still remain questions as to the assignment of critical points.

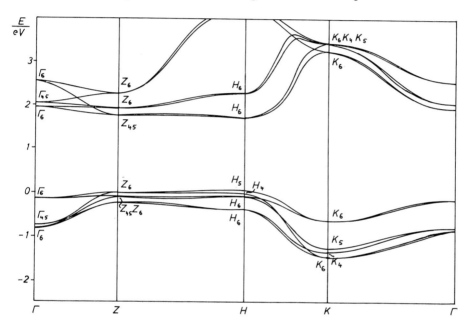

FIG. 4. Tentative band structure of selenium including spin and relativistic corrections.

At any rate the general characteristics of the band structures of selenium are known. Further calculations, however, are needed to obtain the details necessary for a conclusive explanation of all the experimental results known about the properties of these interesting and important semiconductors.

REFERENCES

1. ASENDORF, R. H., *J. Chem. Phys.* **27**, 11 (1957).
2. REITZ, J. R., *Phys. Rev.* **105**, 1233 (1957).
3. OLECHNA, D. J., and KNOX, R. S., *Phys. Rev.* **140**, A986 (1965).
4. BEISSNER, R. E., *Phys. Rev.* **145**, 479 (1966).

5. HULIN, M., *J. Phys. Chem. Solids* **27**, 441 (1966).
6. TREUSCH, J. and SANDROCK, R., *Phys. Status Solidi* **16**, 487 (1966).
7. GROSSE, P. and SELDERS, M., *Phys. Kondens. Materie* **6**, 126 (1967).
8. PICARD, M. and HULIN, M., *Phys. Status Solidi* **23**, 563 (1967).
9. SANDROCK, R., *Phys. Rev.* **169**, 642 (1968); *The Physics of Selenium and Tellurium*, Pergamon, New York, 1969, p. 69.
10. JUNGINGER, H. G., *Solid State Comm.* **5**, 509 (1967).
11. TREUSCH, J., *Phys. Status Solidi* **19**, 603 (1967).
12. KRAMER, B. and THOMAS, P., *Phys. Stat. Sol.* **26** 151 (1968).
13. TREUSCH, J., ECKELT, P. and MADELUNG, O., *Proc. Int. Conf. on II-VI Compounds*, Providence, Rhode Island, U.S.A., 1967.
14. TUTIHASI, S. and CHEN, I., *Phys. Rev.* **158**, 623 (1967).

DISCUSSION

BECKER: Please comment further on the use of a muffin-tin potential for a material like selenium with strong covalent bonding.

MADELUNG: The muffin-tin spheres contact along the chains whereas between the chains a region of constant potential occurs. Thus the directional covalent bond is represented to a certain extent by the arrangement of the spheres. We are well aware of the limitations of this approximation and I have shown at the end of my paper what results of a muffin-tin calculation can be trusted and where the limits of this method lie.

MAGNETOABSORPTION IN TELLURIUM

C. RIGAUX, G. DRILHON and Y. ALPERT

Laboratoire de Physique, Ecole Normale Supérieure, Paris, France

INTRODUCTION

Tellurium is an anisotropic crystal which exhibits dichroism in the fundamental optical absorption.[1] The transition of the electrons across the forbidden gap was investigated by recombination[2] and absorption studies.[3,4] The experimental results indicate that the interband transition is direct and obeys a selection rule: allowed for $E \perp C$, the transition is forbidden for $E \| C$. Theoretical results[5] indicate that this selection rule is valid for the band's extrema located at the points M and P of the Brillouin zone (Fig. 1).

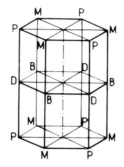

FIG. 1. Brillouin zone of tellurium.

Assuming that the effective mass approximation is valid near the extrema, the isoenergetic surfaces would consist of two revolution ellipsoids around the ternary axis. Results of cyclotron resonance[6] are consistent with such a model. However, evidence for a warped valence band was obtained by Shubnikov–de Haas experiments[7,8] on heavily doped tellurium. In this paper the experimental data obtained in the magnetoabsorption studies performed for both polarizations $E \perp C$ and $E \| C$ are reported. The experiments were carried out at liquid helium temperature on high-purity single crystals of tellurium. An analysis of the results obtained on the allowed and forbidden transitions is performed for the magnetic field applied along the c-axis, and the valence and conduction band parameters are deduced.

EXPERIMENTAL

Material

High-purity single crystals were prepared by a slow cooling of liquid tellurium in a temperature gradient under pure hydrogen atmosphere. Hole concentration and mobility were determined from galvanomagnetic data (Table 1).

TABLE 1. *Hole Concentration and Mobility of Tellurium Single Crystals*

Temperature	77°K	4°K
Hole density (cm^{-3})	10^{14}	10^{13}
Hole mobility (cm^2 V^{-1} sec^{-1})	10^4	10^5

Samples

Different thicknesses and crystalline orientations were used depending on the radiation polarization. For $E \| C$, thick samples (1–5 mm), obtained by cleavage, were optically polished and etched in a chromic solution. By evaporating a $\lambda/4$ layer of selenium, the reflection losses were reduced. For $E \perp C$, the high value of the absorption coefficient requires the use of very thin samples. Films of single crystals (5–10 microns thick) were prepared by grinding plates of tellurium obtained by cleavage and, after optical polish, the samples were etched on both faces to the suitable thickness.

The experimental configurations are described in Fig. 2.

FIG. 2. Experimental configuration. (a) $E \| C$; (b) $E \perp C$.

Experimental Equipment

The experiments were performed at liquid helium temperature in a superconducting magnet producing a field up to 60 kilogauss. The crystals were held on a copper block, the bottom part of which was immersed in liquid helium. Thin samples were mounted loosely between two sapphire plates in order to avoid strains. The experimental set-up is shown in Fig. 3.

The measurements of the transmission were performed at infrared frequencies (energy: 330–370 meV). Using a grating monochromator, the

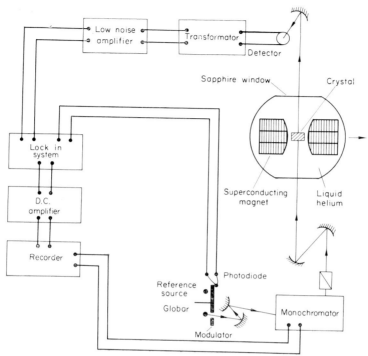

FIG. 3. Experimental set-up.

experimental resolution was about 2.500 in the usual conditions. The monochromatic radiation was focused on the crystal and the light transmitted was concentrated by an elliptical mirror on the detector (InSb p–n junction working at 77°K). The electrical signal was detected by a lock-in amplifier.

RESULTS

Fundamental Absorption

The shape of the absorption edge for both polarizations is indicated in Fig. 4. An exciton peak is observed for $E \| C$. Beyond the continuum limit, the experimental curves $K_\|(\hbar\omega)$ and $K_\perp(\hbar\omega)$ are fitted quantitatively with the theoretical expressions[9] for a direct transition, allowed for $E \perp C$ and forbidden for $E \| C$. If account is taken of the Coulomb interaction:

$$K_{\text{all.}} \propto \frac{\pi\alpha \exp \pi\alpha}{\operatorname{sh} \pi\alpha} \frac{(\hbar\omega - E_g)^{1/2}}{\omega},$$

$$K_{\text{forb.}} \propto \frac{\pi\alpha(1+\alpha^2) \exp \pi\alpha}{\operatorname{sh} \pi\alpha} \frac{(\hbar\omega - E_g)^{3/2}}{\omega},$$

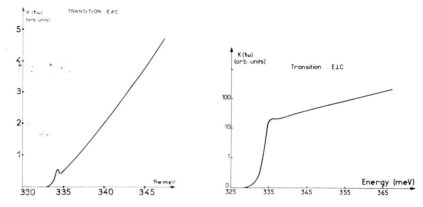

FIG. 4. Absorption coefficient vs. photon energy at 4°K.

where $\alpha^2 = R/(E-E_g)$, E_g = energy gap, and R = exciton binding energy. The comparison between the theoretical and experimental curves is given in Fig. 5 for both polarizations.

R and E_g were deduced from this analysis:

$$E\|C: \quad R = 1.2 \pm 0.2 \text{ meV} \quad E_g = 334.8 \pm 0.2 \text{ meV},$$
$$E\perp C: \quad R = 1 \pm 0.2 \text{ meV} \quad E_g = 335 \pm 1 \text{ meV}.$$

It can be concluded from the absorption edge studies that the transition is direct, allowed for $E\perp C$ and forbidden for $E\|C$. The matrix element $\langle C|\mathbf{p}|V\rangle_{ko}$ is perpendicular to the ternary axis at the extremum $\mathbf{k}_o$ of the bands. Theoretical data obtained by Hulin[5] show that this selection rule is valid for the extremum located at the points M and P of the Brillouin zone.

Magnetoabsorption

Experimental data—$E\|C$. The oscillatory magnetoabsorption is observed at 4°K, even in low magnetic field applied along c. Magnetoabsorption spectra are reported in Figs. 6 and 8. The energies of the peaks are plotted vs. the field H (Fig. 7).

The linear variation for each line (Fig. 7) appears for $H > 20$ kG. The extrapolation at zero field gives the energy gap $E_g = 334$ meV, smaller than the value deduced from the absorption edge data. This fact results from the Coulomb interaction, as the observed oscillations correspond to the transitions into the bound states. In the geometry $E\|C\|H$, a remarkable effect occurs: the intensities distribution depends on the magnetic field direction. This effect appears on pairs of lines, the intensities of which are inverted by the reversal of the magnetic field (Figs. 6 and 8). By performing rotational experiments, it can be concluded that the photon wave vector $\mathbf{q}$ has to be taken into account

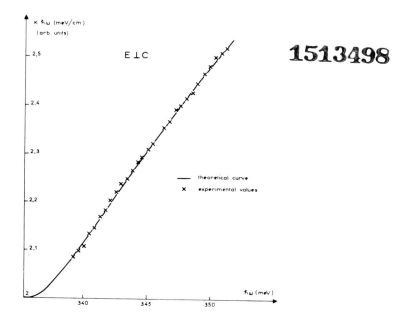

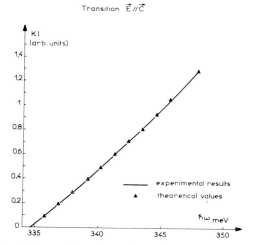

FIG. 5. Comparison of experimental results with theory.

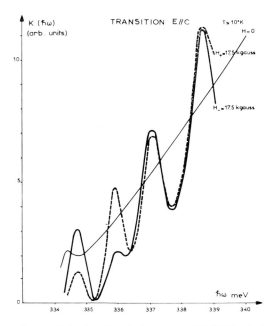

FIG. 6. Absorption coefficient vs. the photon energy $E\|C\|H$ for both directions of magnetic field H^+ and H^- ($H=17.5$ kG).

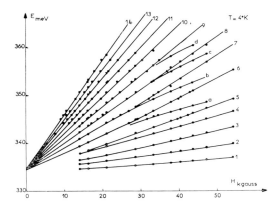

FIG. 7. Energy of the peaks vs. the field $E\|C\|H$.

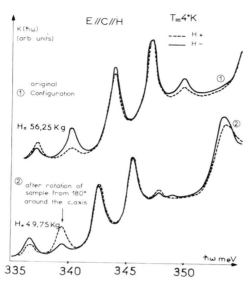

FIG. 8. Magnetoabsorption spectra. 1. For both directions of magnetic field H^+ and H^- in the original configuration described by Fig. 9a. 2. After a rotation of the sample 180° around c, corresponding to the configuration described by Fig. 9c.

to interpret this effect. A rotation of the crystal (in Fig. 9a) from 180° around the c-axis was performed (Fig. 9c). The initial configuration is equivalent to the situation (Fig. 9b) resulting from a rotation of the whole system (crystal magnetic field, radiation) from 180° around the direction 3. It is observed that the rotation of the crystal is equivalent to the reversal of **q** in the first situation (Figs. 9b and 9c). Figure 8 shows the oscillatory spectra for both directions of the field H^+ and H^- in the original configuration and after rotation of the sample. This experiment demonstrates that the reversal of **q** into −**q** for a

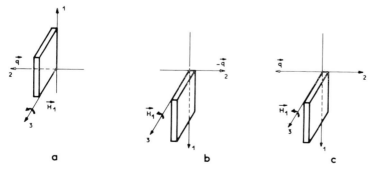

FIG. 9. Experiment involving the rotation of the sample about the c-axis.

given direction of **H** changes one type of spectra into the other, and produces the same result as the reversal of the field for a given direction of **q**.

Experimental data—$E \perp C$. In the case of the allowed transition, oscillations of the absorption coefficient appear for magnetic fields larger than 25 kG. Oscillatory spectra and the energies of the peaks vs. the field are indicated in Figs. 10 and 11.

A comparison of the results obtained for the allowed and forbidden transitions shows that the peaks occur at the same energies in both cases. However, the intensities of the lines are quite different for $E \| C$ and $E \perp C$. Complementary spectra are obtained, the stronger oscillations for one polarization corresponding to the weaker for the other.

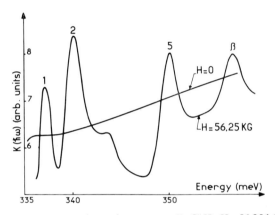

FIG. 10. Magnetoabsorption spectra. $E \perp C \| H$; $H = 56.25$ kG.

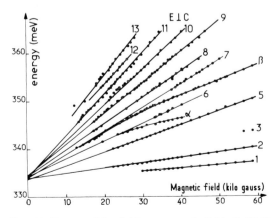

FIG. 11. Energy of the field vs. magnetic field. $E \perp C \| H$.

Theoretical data. The experimental results have been analyzed in the framework of the effective mass theory. If one assumes a quadratic energy-momentum relation near the extremum, the isoenergetic surfaces consist of two revolution ellipsoids around the ternary axis, one for the six equivalent points M and one for the six equivalent points P, deduced one from the other by the symmetry with respect to the center of the zone. These two ellipsoids correspond to the same energy, but to different spin states. Before discussing the results obtained in tellurium, some theoretical aspects of the absorption edge in a magnetic field will be considered.

The interaction of the radiation with the electrons is described by the time-varying perturbation

$$\mathcal{H}_I = \frac{eA_0}{mc} \boldsymbol{\epsilon} \, e^{i(\mathbf{q}\cdot\mathbf{x}-\omega t)} \cdot \boldsymbol{\pi}$$

in the hamiltonian. $\boldsymbol{\pi}$ is the operator $\mathbf{p}+e\mathbf{A}/c$, where $\mathbf{A}$ is the vector potential due to the magnetic field. The photon wave vector $\mathbf{q}$ is along the x-axis. A_0 is the amplitude of the vector potential due to the electromagnetic field. Using the semi-classical radiation theory, the absorption coefficient K is deduced from the transition probability for the process in which an electron is excited from the band 1 to the band 2 by a photon absorption. For direct transitions[10, 11]

$$K(\hbar\omega) = \frac{4\pi^2 e^2}{m^2 \omega \eta c} \sum_{1,2} \langle 2 | \boldsymbol{\epsilon}\, e^{i\mathbf{q}\cdot\mathbf{x}} \cdot \boldsymbol{\pi} | 1 \rangle \delta(E_2 - E_1 - \hbar\omega) \quad (1)$$

(η = refractive index)

the sum is over initial and final states paired by the selection rules. We consider only the case of ellipsoidal bands having a revolution axis parallel to the field $H\|z$. The energies of the electrons in the two bands are given by:

$$E_1 = E_{10} - (n_1 + \tfrac{1}{2})\hbar\omega_1 - \frac{\hbar^2 k_z^2}{2m_{1l}} \pm \tfrac{1}{2} g_1 \mu_B H \quad \text{(valence)},$$

$$E_2 = E_{20} + (n_2 + \tfrac{1}{2})\hbar\omega_2 + \frac{\hbar^2 k_z^2}{2m_{2l}} \pm \tfrac{1}{2} g_2 \mu_B H \quad \text{(conduction)},$$

$$\omega_1 = \frac{eH}{cm_{1t}}; \quad \omega_2 = \frac{eH}{cm_{2t}};$$

m_{1l}, m_{2l}: longitudinal effective masses,

m_{1t}, m_{2t}: transverse effective masses.

The matrix element to the first order in $\mathbf{q}$ is given by:

$$\langle 2 | \boldsymbol{\epsilon}\, e^{i\mathbf{q}\cdot\mathbf{x}} \cdot \boldsymbol{\pi} | 1 \rangle = \delta(\mathbf{K}_2 - \mathbf{K}_1 - \mathbf{q}) \left\{ \boldsymbol{\epsilon}\cdot\mathbf{p}_{12} F_N(0) + [\mathbf{M}\cdot\boldsymbol{\pi} F_N(r)]_{r=0} + \frac{\hbar}{2} \mathbf{q}\cdot\boldsymbol{\mu} F_N(0) \right.$$
$$\left. + \frac{\hbar}{mE_g}[\mathbf{q}\cdot\mathbf{p} F_N(r)]_{r=0} \right\} \quad (2)$$

which implies the conservation of the photon and electron momenta: $\mathbf{q}=\mathbf{K}_2-\mathbf{K}_1$. $F_N(r)$ is the wave function of the relative motion of the electron and hole and r the relative distance.

$$\mathbf{p}_{12}=\frac{(2\pi)^3}{\Omega}\int_{\text{cell}} u_{20}^*(\tau)\mathbf{p}u_{10}(\tau)\,d\tau, \tag{3}$$

$$\mathbf{M}=\frac{1}{m}\left[\sum_{\alpha\neq 2}\frac{\mathbf{p}_{\alpha 2}(\boldsymbol{\epsilon}\cdot\mathbf{p}_{1\alpha})}{E_2-E_\alpha}+\sum_{\alpha\neq 1}\frac{\mathbf{p}_{1\alpha}(\boldsymbol{\epsilon}\cdot\mathbf{p}_{\alpha 2})}{E_1-E_\alpha}\right], \tag{4}$$

$$\boldsymbol{\mu}=\frac{1}{m}\left[\sum_{\alpha\neq 2}\frac{\mathbf{p}_{\alpha 2}(\boldsymbol{\epsilon}\cdot\mathbf{p}_{1\alpha})}{E_2-E_\alpha}-\sum_{\alpha\neq 1}\frac{\mathbf{p}_{1\alpha}(\boldsymbol{\epsilon}\cdot\mathbf{p}_{\alpha 2})}{E_1-E_\alpha}\right]. \tag{5}$$

In the dipolar approximation, the wavelength of light is considered to be infinite and q is neglected. The two first terms in expression (2) describe the electric dipole transitions. Light of finite wavelength interacting with the electrons can permit other transitions; q is small compared to the reciprocal lattice vector, and the linear term in q is considered in the matrix element (2).

In the dipolar approximation, the selection rules and expressions for the absorption coefficient[11] are summarized as follows (Table 2):

TABLE 2. *Absorption Coefficient Expressions*

	Selection rules	Energies of the transitions	$K(\hbar\omega)$
Allowed transition	$n_1=n_2$	$E_n=E_g+(n+\tfrac{1}{2})\hbar\omega^*$	$\dfrac{e^3 H(2\mu)^{1/2}}{2m^2\omega\eta c\hbar^2}\|\boldsymbol{\epsilon}\cdot\mathbf{p}_{12}\|^2 \times \sum_n (\hbar\omega-E_n)^{-1/2}$
Forbidden transition $M\|H$	$n_1=n_2$	$E_n=E_g+(n+\tfrac{1}{2})\hbar\omega^*$	$\dfrac{e^3 H(2m^*)^{3/2}}{2m^2\omega\eta c^2\hbar^2}(M_z)^2 \times \sum_n (\hbar\omega-E_n)^{+1/2}$
$M\|H$	$n_2=n_1+1$	$E_{n_1}=E_g+(n+\tfrac{1}{2})\hbar\omega^*+\hbar\omega_1$	$\dfrac{e^4 H^2(2\mu)^{1/2}}{4m^2 c^3\hbar\omega\eta}\|M^2\|(n+1)$
	$n_1=n_2-1$	$E_{n_2}=E_g+(n+\tfrac{1}{2})\hbar\omega^*+\hbar\omega_2$	$\times\sum_n\{(\hbar\omega-E_{n_1})^{-1/2}+(\hbar\omega-E_{n_2})^{-1/2}\}$

Taking into account the spin splitting, each series of lines indicated above is split into two components separated by $\tfrac{1}{2}(g_1+g_2)\mu_B H$ or $\tfrac{1}{2}(g_1-g_2)\mu_B H$ depending on the light polarization.

DISCUSSION

The experimental results obtained for $E \perp C$ and $E \| C$ were compared with the theoretical data. The analysis was performed in the high field range 25 kG $< H <$ 40 kG, where the Coulomb interaction introduces only a small perturbation on the magnetoabsorption effects ($\hbar\omega^* \gg 2R$).

A periodic repartition in the energies of the peaks for both transitions is observed, the separation between the peaks such as (2–7–11; 5–9–13; 6–10; 8–12) is equal to $\hbar\omega^*$. For $H = 37.5$ kG, $\hbar\omega^* = 14.5 \pm 0.2$ meV, and one can deduce the reduced transverse effective mass, $m^* = 0.029\, m_0$.

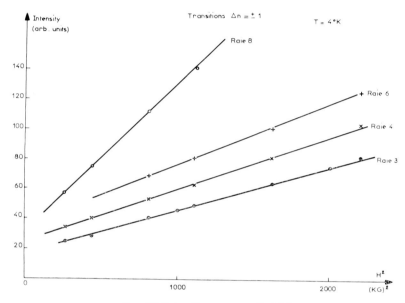

FIG. 12. Intensity vs. magnetic field.

The transitions $\Delta N = 0$ and $\Delta N = \pm 1$ are selected by comparing the experimental data obtained on the allowed and forbidden transitions. For $E \| C$, the oscillations are observed for $H \| C$. This fact is in agreement with the theory, as the symmetry considerations permit the deduction that the **M** vector is perpendicular to the ternary axis and that $\mathbf{M} \perp \mathbf{H}$. In this geometry, the transitions $\Delta N = \pm 1$ should be observed. Two series of strong peaks (3, 4, 6, 8, 10, 12) would correspond to these transitions. The intensities increase as the square of the field (Fig. 12) as predicted by the theory. A further confirmation is provided by the results obtained on the allowed transition $E \perp C$ as these lines are missing or very weak for this polarization. In the geometry $E \| C$, additional series of peaks occur, the intensity of which depends on the photon wave vector direction. The energies of these transitions correspond exactly to the stronger oscillations in the allowed case $E \perp C$ which are attributed to the $\Delta N=0$ transitions. Taking into account the finite wavelength of light, the existence of these lines and the modification of the intensity resulting from the reversal of **q** is understood, if we assume that the selection rule ($p''_{12}=0$) is not very strict. For small p''_{12}, the absorption coefficient corresponding to the $\Delta N=0$ is proportional to $p''_{12}(p''_{12}+\hbar \mathbf{q} \cdot \boldsymbol{\mu})$, where $\boldsymbol{\mu}$ is given by expression (5). The change of **q** to $-\mathbf{q}$ modifies the intensity of the peak into: $p''_{12}(p''_{12}-\hbar \mathbf{q} \cdot \boldsymbol{\mu})$. For such transitions the variation of the intensity should depend linearly upon H. This fact is verified by the experimental results (Fig. 13).

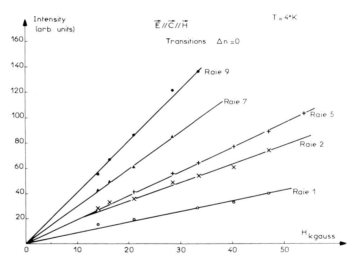

FIG. 13. Intensity vs. magnetic field.

Determination of Band Parameters

For a quantitative interpretation of the energies $E_n(H)$ as a function of the field, the spin splitting of the Landau levels has to be taken into account. For the first transition (0 0), the observed variation $E(H)$ corresponds quantitatively to the theoretical expression

$$E = Eg + \tfrac{1}{2}\frac{eH\hbar}{m^*c} \pm \tfrac{1}{2}\gamma \mu_B H,$$

with the values previously deduced for E_g and m^*. $\gamma = g_1 + g_2$, or $g_2 - g_1$. The complete energy spectrum is computed for the allowed and forbidden cases, and the band parameters are fitted by comparison with the experimental data.

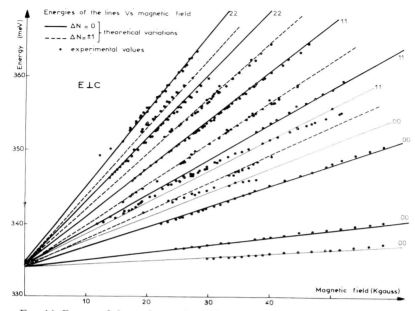

FIG. 14. Energy of the peaks vs. the field ($E \perp C$). Theoretical and experimental variations.

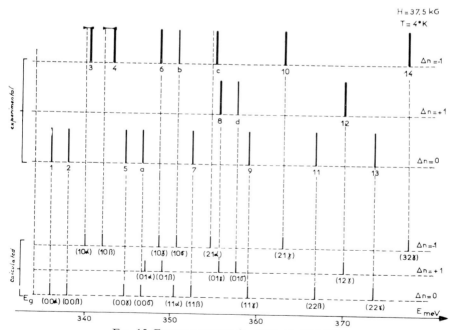

FIG. 15. Energy spectrum for $E \parallel C$. $H = 37.5$ kG.

Figure 14 shows the observed and calculated variations $E_n(H)$ for $E \perp C$. The energy spectrum for $E \| C$ is indicated in Fig. 15 for $H = 37.5$ kG. The four possible transitions between the spin states seem to be observed for $E \| C$ and $E \perp C$. The band parameters deduced by this analysis are: $m_{1t} = (0.113 \pm 0.006)m_o$ in good agreement with the hole transverse effective mass obtained by cyclotron resonance; $m_{2t} = (0.039 \pm 0.001)m_o$; $g_1 = 10 \pm 1$; $g_2 = 35 \pm 4$; $E_g = 334$ meV.

CONCLUSION

Most of the experimental results obtained in the magnetoabsorption studies on the transitions $E \perp C$ and $E \| C$ for $H \| C$ can be interpreted by the effective mass theory. The transition (334 meV) is direct, but the selection rule is probably not very strict, as indicated by the observation of the $\Delta N = 0$ transitions in the forbidden case. Thus, it seems possible that the extremum of the bands would be located not exactly in M and P but in the vicinity of these points. The influence of the photon wave vector in such transitions has to be considered to explain the modification of the intensity distribution which occurs with the reversal of the photon wave vector in the geometry $E \| C$. The data obtained (variations of the energies and intensities of the peaks with magnetic field, influence of the reversal of $\mathbf{q}$ on the intensities) are interpreted for the transition at 334 meV and the band parameters are deduced. However, in the presence of higher fields ($H > 40$ kG), another series of transitions (Figs. 7 and 11) has been observed for $E \| C$ and $E \perp C$ which cannot be explained in this analysis. The existence of another transition at higher energy (about 338 meV) may be invoked to understand the presence of such lines. At the present time, a definite conclusion cannot be drawn and more complete results have to be obtained.

REFERENCES

1. BLAKEMORE, J. S. and NOMURA, K. C., *Phys. Rev.* **127**, 1024 (1962).
2. BENOIT À LA GUILLAUME, C. and DEBEVER, J. M., *Proc. Conf. Physics on Quantum Electronics*, 1966, p. 397.
3. RIGAUX, C. and DRILHON, G., *Proc. Int. Conf. Physics of Semiconductors*, Kyoto, The Physical Society of Japan, Tokyo, 1966, p. 193.
4. ALPERT, Y. and RIGAUX, C., *Solid State Comm.* **5** (5), 391 (1967).
5. HULIN, M., *J. Phys. Chem. Solids*, **27**, 441 (1966).
6. PICARD, J. C. and CARTER, D., *Proc. Int. Conf. Physics of Semiconductors*, Kyoto, The Physical Society of Japan, Tokyo, 1966, p. 202.
7. GUTHMANN, C. and THUILLIER, J. M., *Compt. Rend.* **263B**, 303 (1966).
8. BRAUN, E. and LANDWEHR, G., *Proc. Int. Conf. Physics of Semiconductors*, Kyoto, The Physical Society of Japan, Tokyo, 1966, p. 380.
9. MCLEAN, T. P., *Progress in Semiconductors* **5**, 55 (1960).
10. ROTH, L. M., LAX, B. and ZWERDLING, S., *Phys. Rev.* **114**, 90 (1959).
11. ELLIOTT, R. S., MCLEAN, T. P. and MCFARLANE, G. G., *Proc. Phys. Soc.*, **72**, 553 (1958).

DISCUSSION

BECKER: Please give more information about the new lines which you mentioned for $H > 40$ kG.

RIGAUX: For magnetic fields larger than 40 kG, we observe another series of lines for both polarizations $E \| C$ and $E \perp C$. The existence of another transition at about 338 meV may be invoked to explain this feature. More complete experimental results have to be obtained to understand these data.

BECKER: Have you performed any similar studies on selenium?

RIGAUX: No, we studied only tellurium.

SHUBNIKOV–DE HAAS EFFECT IN TELLURIUM

C. GUTHMANN and J. M. THUILLIER

Laboratoire de Physique, Ecole Normale Supérieure, Paris, France

QUANTIZATION of the orbital motion of electrons in a magnetic field gives rise to a quasi-periodic variation of the density of states as a function of energy. Consequently, transport phenomena show an oscillatory behavior as a function of magnetic field. The period of the oscillations is given by $\Delta H^{-1} = (2\pi e/\hbar c A)$, where A is the extremal cross-section of the Fermi surface in a plane perpendicular to the magnetic field, and the other constants have their usual meaning. In tellurium which is doped heavily enough to exhibit p-type conduction at very low temperatures, it is possible to observe the Shubnikov–de Haas effect and hence study the constant energy surfaces of the valence band.

As well as having sufficient ionized carriers at low temperatures, the necessary conditions for the observation of oscillatory magnetoresistance are: $\zeta_0 > kT$ (degenerate material), $\zeta_0 > \hbar\omega_c$, and $\omega_c \tau > 1$, where $\omega_c = eH/m^*c$ is the cyclotron frequency, τ the collision time, and ζ_0 the Fermi energy in the absence of magnetic field. The last inequality implies a material with a high mobility.

These conditions can be fulfilled with doped material (10^{17} to 2×10^{18} cm^{-3}) in a strong magnetic field at liquid helium temperature.

In order to provide the large magnetic fields required, a pulsed magnetic field is used which can produce during some milliseconds fields up to 150 kG in a transverse geometry, and up to 280 kG in a longitudinal one.

The rapid variation of H induces a voltage in the sample, which must be eliminated. A pulse of a.c. current (500 kHz) is triggered when the magnetic field reaches its maximum value, and the measurements are made as H decreases. The induction voltage is at low frequency, and can easily be separated from the magnetoresistance voltage.

The samples are cut along the c-axis or along a binary axis. The oscillatory magnetoresistance is investigated in both longitudinal and transverse magnetic fields as a function of the angle between the c-axis or binary axis and the magnetic field.

EXPERIMENTAL RESULTS

Samples having a hole concentration of 6×10^{17} cm^{-3} and 1.8×10^{18} cm^{-3} were studied in detail.

For samples cut along the c-axis no change was observed in the periodicity of the oscillations when the magnetic field was rotated in a plane perpendicular to this axis.

It can be concluded that the constant energy surfaces have rotational symmetry around the c-axis.

For samples cut along a binary axis, the situation is more complicated. The direction of the magnetic field is shown in Fig. 1. Let θ be the angle between H and the c-axis.

For $\pi/6 < \theta < \pi/2$, the oscillations are periodic (Fig. 2), and the extremal areas of the Fermi surface for different values of θ (Fig. 3) can be deduced.

For $\theta < \pi/6$, it was not possible to analyze the experimental results with a single period structure (Fig. 2). Braun and Landwehr[1] obtained similar results.

The position of the successive maxima and minima of magnetoresistance cannot be explained by spin splitting.

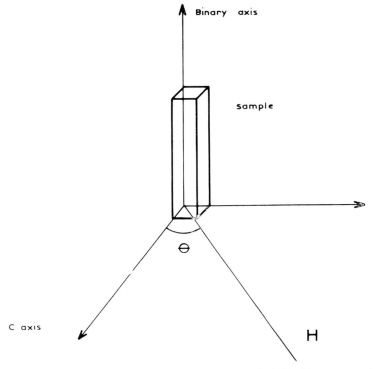

FIG. 1. Diagram showing the orientation of the magnetic field with respect to the c-axis in the sample.

SHUBNIKOV–DE HAAS EFFECT IN TELLURIUM 49

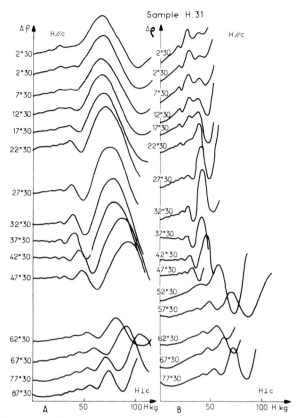

FIG. 2. Variation of the resistance change $\Delta\rho$ with magnetic field strength for various angles θ between H and the c-axis. A: Arbitrary ordinate scale. B: Ordinate scale 5 times that of A.

However, the position and the amplitude of the oscillations can be explained by a double period model, where the ratio of the two corresponding extremal areas depends on the orientation. This ratio varies from 1 for $\theta = \pi/6$ to 1.8 for $H \| C$ (Fig. 3). The experimental results for two typical samples are summarized in Table 1.

TABLE 1. *Experimental Results for Typical Tellurium Samples*

Sample		$H \perp C$		$H \| C$	
H 31 $p = 6 \times 10^{17}$ cm^{-3}	period g^{-1} extremal areas cm^{-2}	0.5×10^{-5} 2×10^{13}	1×10^{-5} 10^{13}	1.8×10^{-5} 0.55×10^{13}	
T '25 $p = 1.8 \times 10^{18}$ cm^{-3}	period g^{-1} extremal areas cm^{-2}	0.26×10^{-5} 3.8×10^{13}	0.55×10^{-5} 1.8×10^{13}	0.9×10^{-5} 1.1×10^{13}	

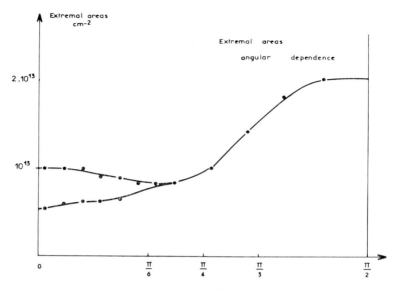

Fig. 3. Plot of extremal areas of the Fermi surface against angle θ.

DISCUSSION

From the experimental results a symmetry of revolution around the c-axis was deduced. However, the results do not exclude a ternary warping of the Fermi surface.

In the Shubnikov–de Haas effect a ternary warping should be observed with a six-fold symmetry, and the corresponding variations of the extremal cross-sections should be rather weak.

The angular dependence of the period of the oscillations permits the approximate determination of the Fermi surface of the valence band: it is a dumbbell shaped surface (Fig. 4).

In the band model given by Picard and Hulin[2] the maximum of the valence band lies on the edge of the Brillouin zone, near a corner. This model of the Fermi surface corresponds to two pairs of ellipsoids for pure samples and low temperature. For higher carrier concentration each pair coalesces to give a single surface. This theory can explain the constriction of the Fermi surface at the central section.

This model explains also the difference between the present results and the cyclotron resonance results.[3]

Material in a more extended range of carrier concentration must now be studied to try to see a possible deformation of the Fermi surface with the hole concentration.

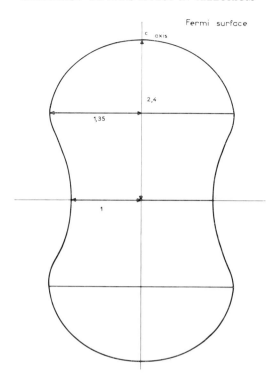

FIG. 4. Calculated Fermi surface of the valence band showing the dumbbell shape.

ACKNOWLEDGMENT

The authors wish to thank Professor Hulin for very enlightening discussions.

REFERENCES

1. BRAUN, E. and LANDWEHR, G., *Proc. Int. Conf. Physics of Semiconductors*, Kyoto, The Physical Society of Japan, Tokyo, 1966, p. 380.
2. PICARD, M. and HULIN, M., *Phys. Status Solidi* **23**, 563 (1967).
3. PICARD, J. C. and CARTER, D. L., *Proc. Int. Conf. Physics of Semiconductors*, Kyoto, The Physical Society of Japan, Tokyo, 1966, p. 202.

DISCUSSION

CHAMPNESS: What was the duration of the magnetic pulse in your experiment and how was the measurement carried out?

GUTHMANN: The rise time of the magnetic field is 3 milliseconds. The measurements were done during the decrease of the magnetic field.

STUKE: Braun and Landwehr[1] observed an additional structure in the form of the Fermi surface. What is the reason for this difference?

GUTHMANN: I observed the same experimental curves as Braun and Landwehr. A double period model where H is near the c-axis, which gives a dumbbell-shaped Fermi surface, is more consistent with the valence band model and the cyclotron resonance results.

BECKER: Have you measured or analyzed the damping term in the Shubnikov–de Haas effect?

GUTHMANN: No, I have not yet made precise amplitude measurements as a function of temperature.

HEDGCOCK: Did you observe a negative magnetoresistance in any of the tellurium samples that you studied?

GUTHMANN: It was not possible to observe the effect from our experiments.

INTERBAND TRANSITION OF HOLES IN TELLURIUM

D. HARDY and C. RIGAUX

Laboratoire de Physique, Ecole Normale Supérieure, Paris, France

INTRODUCTION

The existence of a strong infrared absorption, centred at 11 microns and only observed for the radiation polarization parallel to the c-axis, was reported previously by Caldwell and Fan.[1] The results obtained in the infrared absorption experiments [1, 2] on various doped crystals, in the temperature range 100–300°K, show that the intensity absorbed is proportional to the density of the holes, which implies that this absorption corresponds to an interband transition of the holes. Theoretical data on the band structure explain the existence of such a transition and the observed selection rule, if the spin-orbit coupling is taken into account. From Hulin's [3, 4] theoretical results, the transition is thought to take place between the two higher valence bands separated by about 0.1 eV, at the corners of the Brillouin zone or near these points. The transition is allowed for $E \| C$ and forbidden for $E \perp C$ in agreement with experimental observation.

The infrared absorption at 11 microns was investigated in order to state precisely the valence band structure near the maximum. Transmission measurements were carried out at low temperature, on low doped single crystals of tellurium. The variations of the absorption coefficient $K(\hbar\omega)$ were determined as a function of the photon energy, in the range 100–150 meV, for light polarizations $E \| C$ and $E \perp C$. In addition, the interband transition was investigated in the presence of a magnetic field at low temperature.

EXPERIMENTAL

Variously doped samples of crystals of tellurium were prepared for these experiments. The hole density p and the mobility μ were determined from galvanomagnetic data: at 77°K, $p = 10^{14}$ to 5×10^{15} cm^{-3} and $\mu = 7 \times 10^3$ to 10^4 cm^2 V^{-1} sec^{-1}. The experimental set-up was similar to the equipment used in the magnetoabsorption studies:[5] the detector was a photoconducting Ge–Hg crystal cooled at 20°K. Using a grating monochromator, the experimental resolution was about 1000 in the spectral range 8–13 microns.

RESULTS

Absorption E∥C

The absorption curves $K(\hbar\omega)$ obtained in the geometry $E\|C$, at low temperature (4°K, 20°K, 77°K) are indicated in Figs. 1, 2 and 3. At 4°K a structure is observed in the absorption curve which appears for the low doped crystals ($p = 10^{14}$ to 7×10^{14} cm^{-3}). The spectrum consists of a broad absorption band centred at $E_1 = 126.3$ meV and a peak located at $E_2 = 128.6$ meV.

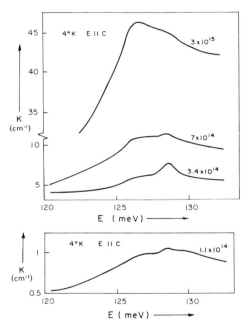

FIG. 1. Variation of absorption coefficient with photon energy at 4°K for samples with different hole concentrations.

This structure disappeared by increasing the dopage or the temperature. At 20°K and 77°K, the absorption was broadened in the lower energy transition and the high energy peak became indistinguishable.

At 20°K, $E_1 = 125.9$ meV.

At 77°K, $E_1 = 123.7$ meV.

Magnetoabsorption

Preliminary results were obtained in the magnetoabsorption studies on the hole interband transition. At 4°K, in the presence of a magnetic field ($H < 20$ kG) applied along the ternary axis, no effect was observed. However,

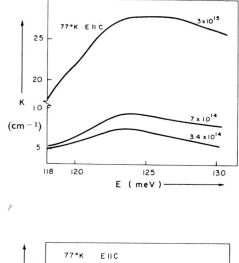

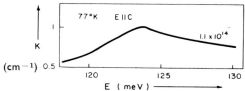

FIG. 2. Variation of absorption coefficient with photon energy at 77°K for samples with different hole concentrations

for a magnetic field applied along a binary axis, the two absorptions observed at zero field were shifted towards the higher energies and an oscillatory spectrum appeared (Fig. 4).

For the higher energy transition, the energies of the peaks depend linearly upon the magnetic field (Fig. 5), and the extrapolation at zero field of the energies corresponds to the observed oscillation $E_2 = 128.6$ meV.

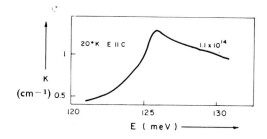

FIG. 3. Variation of absorption coefficient with photon energy at 20°K for a sample with a hole concentration of 1.1×10^{14} cm^{-3}.

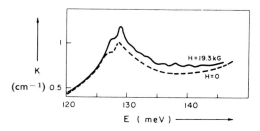

FIG. 4. Variation of absorption coefficient with photon energy at 4°K in a magnetic field of 0 and 19.3 kG applied along a binary axis.

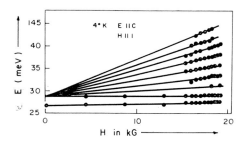

FIG. 5. Plot of the energies of the absorption peaks against magnetic field strength.

For the lower energy transition, the variation $E(H)$ is non-linear, but involves a quadratic term.

DISCUSSION

The experimental data reported above (structure in the absorption curve observed at 4°K and magnetic behavior) imply the existence of two transitions separated by 2.3 meV at 4°K, which obey a selection rule (allowed for $E \| C$ and forbidden for $E \perp C$). From theoretical data on band structure calculations performed by Hulin [3, 4] (Fig. 6), the transition is thought to take place

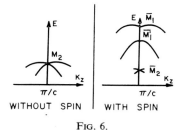

FIG. 6.

between the two higher levels of valence band $\overline{M}_1$ and $\overline{M}'_1$ of the four bands arising from the introduction of spin effects on the M_2 crossed bands. These two levels $\overline{M}_1$ and $\overline{M}'_1$ culminate at or near the corners of the Brillouin zone, where the selection rule $\overline{M}_1 \rightarrow \overline{M}'_1$ corresponds to the experimental observation (no peak in the absorption curve was observed for $E \perp C$). Evidence for a warped valence band has been provided by cyclotron resonance[6] and Shubnikov–de Haas effect[7] experiments. In addition, magnetoabsorption[5] results indicate that the extremum of the valence band should probably be located in the vicinity of the zone corners.

FIG. 7.

An attempt to interpret the structure observed at 4°K in the 11-micron absorption $(E \| C)$ was carried out, using a warped valence band model. Assuming the extremum of the valence band (before introducing spin-orbit interaction) located near the zone corners of the Brillouin zone, the isoenergetic surfaces would be equivalent to two revolution ellipsoids around the ternary axis

$$E_1(k) = -a(k_z - k_0)^2 - bk_\perp^2,$$
$$E_2(h) = -a(k_z + k_0)^2 - bk_\perp^2.$$

Introducing spin-orbit interaction, the energies of the two levels $\overline{M}_1$ and $\overline{M}'_1$ are given by (Fig. 7):

$$E^\pm = -a(k_z^2 + k_0^2) - bk_\perp^2 \pm \sqrt{(\lambda^2 + 4a^2 k_0^2 k_z^2)}$$

In this simplified model, the absorption coefficient given by

$$K(\hbar\omega) \propto \frac{\hbar\omega}{\sqrt{[(\hbar\omega)^2 - 4\lambda^2]}} \exp -\{(\hbar\omega - \Delta)^2 / 4\Delta kT\} \text{ (for direct allowed transitions)}$$

explains the existence of two absorption peaks for $\hbar\omega = 2\lambda$ on the axis and for $\hbar\omega = \Delta$ at the maximum of the upper band. However, the calculated $K(\hbar\omega)$ disagrees with the experimental absorption shape for the lower energy transition.

More complete analysis is at present being carried out on the interpretation of the observed features.

REFERENCES

1. CALDWELL, R. S. and FAN, Y. H., *Phys. Rev.* **114**, 664 (1959).
2. GROSSE, P. and SELDERS, M., *Phys. Kondens, Materie* **6**, 126 (1967).
3. HULIN, M., *J. Phys. Chem. Solids* **27**, 441 (1966).
4. PICARD, M. and HULIN, M., *Phys. Status Solidi* **23**, 563 (1967).
5. RIGAUX, C., DRILHON, G. and ALPERT, Y., *The Physics of Selenium and Tellurium*, Pergamon, New York, 1969, p. 31.
6. PICARD, J. C. and CARTER, D., *Proc. Int. Conf. Physics of Semiconductors*, Kyoto, The Physical Society of Japan, Tokyo, 1966, p. 202.
7. GUTHMANN, C. and THUILLIER, J. M., *The Physics of Selenium and Tellurium*, Pergamon, New York, 1969, p. 47.

TRAPPING LEVELS IN HEXAGONAL SELENIUM

H. P. D. Lanyon[†] and R. M. Krambeck[‡]

Department of Electrical Engineering, Carnegie-Mellon University, Pittsburgh, Pa. 15213

INTRODUCTION

In this paper measurements are described of the a.c. capacitance of selenium single crystals with a blocking contact and of the resistance in series with the barrier capacitance. Measurements were made using metal contacts evaporated onto the natural ($10\bar{1}0$) cleavage planes of the selenium crystals. The crystals used were both pure selenium[§] and selenium doped with potassium or thallium.[††] Evaporated nickel electrodes gave ohmic contact to the selenium, evaporated aluminum electrodes were used for the blocking contact. The band model and equivalent circuit of the sample are shown in Fig. 1.

In the first part of the paper, measurements are described of the impedance of the samples using a GR 1608A impedance bridge. The voltage dependence of the surface capacitance is close to that of a Schottky barrier. This voltage dependence has been measured as a function of temperature. An irreversible change in the acceptor concentration upon heating suggests that an annealing of defects occurs above room temperature. The series resistance of the samples was substantially independent of voltage and little dependent on temperature indicating that the nickel provides a good ohmic contact and that the resistivity of the selenium is essentially temperature independent.

In the second part of the paper, measurements are described of the changes in capacitance seen upon illumination of the sample under zero bias. Carriers excited by the light are swept out of the barrier region by the dipole field changing the effective acceptor concentration and hence the capacitance. Upon removal of the illumination the charge distribution decays back to its thermal equilibrium state. In contrast to the previous measurements the decay rate is strongly temperature dependent, being controlled by the depth of the energy level from which the free carriers were originally excited. Decay

---

[†] Present address: Department of Electrical Engineering, Worcester Polytechnic Institute, Worcester, Mass. 01609.
[‡] National Science Foundation Fellow.
[§] By courtesy of J. D. Harrison and D. E. Harrison, Westinghouse Electric Corporation.
[††] By courtesy of R. C. Keezer, Xerox Corporation.

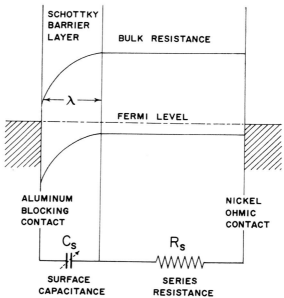

Fig. 1. Equivalent circuit of sample with one blocking and one ohmic contact.

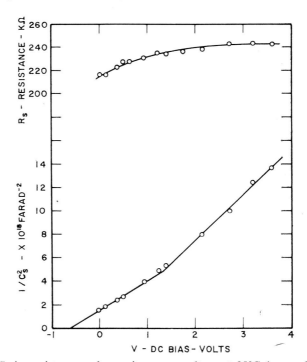

Fig. 2. Series resistance and capacitance vs. voltage at 25°C (pure selenium).

times (to 1/e of the original change) from 0.3 to 1000 seconds have been measured using a PAR Model HR–8 lock-in amplifier to monitor the capacitance changes. The levels involved are 0.7 eV above the valence band in the pure and thallium-doped samples; 0.5 eV in the potassium-doped samples.

STEADY STATE MEASUREMENTS

It can be shown [1] that for a Schottky barrier the surface capacitance is related to the doping content N_a and the applied voltage V through the relationship $1/C_s^2 = 2(V_B+V)/\epsilon\epsilon_0 N_a q$ where V_B is the initial barrier height and $\epsilon\epsilon_0$ is the dielectric constant of the semiconductor. Figure 2 is a typical plot of $1/C_s^2$ vs. V for a pure selenium sample taken at 1 kHz; the polarity is that of the aluminum with respect to the nickel electrode. Two straight-line plots (which would be expected for a Schottky barrier) have been drawn. From the slopes one can estimate an acceptor concentration of approximately $4 \times 10^{14}/cm^3$ at 2 microns from the interface, increasing to $5.7 \times 10^{14}/cm^3$

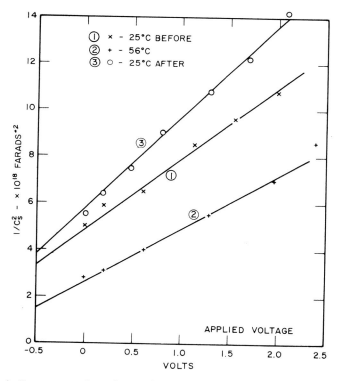

FIG. 3. Temperature dependence of capacitance in thallium-doped selenium at 1 kHz.

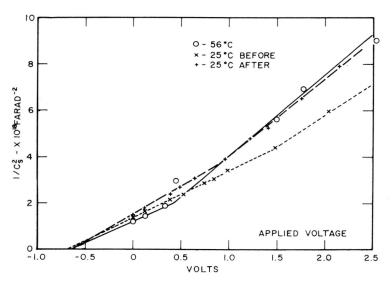

Fig. 4. Capacitance vs. d.c. voltage for pure selenium at 1 kHz.

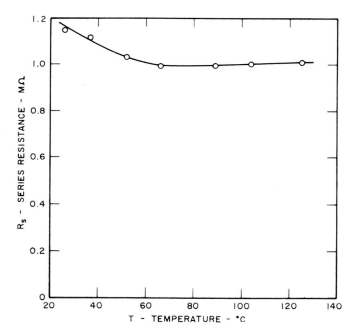

Fig. 5. Temperature dependence of series resistance in thallium-doped selenium.

within 1 micron of the surface. These values were reproducible on both cleaved and etched surfaces. The electrode could be removed and the sample etched in Na_2S without a drastic change in the measured acceptor concentration. This suggests that there is a maximum of 4×10^{14} such centres/cm^3 in the bulk of the material. The intercept on the voltage axis gives a value of $V_B = 0.6^v$ for the initial barrier height at the aluminum/selenium interface. Similar results are obtained for samples doped with potassium or thallium. The voltage dependence of the series resistance is also shown in Fig. 2. There is only 10 per cent change in the voltage range covered.

The effect of heating the samples is shown in Figs. 3 and 4. Figure 3 shows the capacitance of a thallium-doped sample before and after heating to 56°C and at 56°C. Upon heating the sample, the apparent acceptor concentration increased from 5×10^{14}/cm^3 to 7×10^{14}/cm^3. However, on recooling to 25°C the acceptor concentration fell to 4×10^{14}/cm^3 (which is below the original concentration). After the initial heating the sample could be recycled without further change provided the previous maximum temperature was not exceeded. On increasing the temperature further, an additional change in acceptor concentration was observed. Equivalent changes were observed when the samples were heated in air or *in vacuo* and with doped and undoped samples. The behavior of an undoped sample is shown in Fig. 4. It is concluded that the process is caused by the annealing of lattice defects rather than by the oxidation or diffusion of dopants in the selenium.

The temperature dependence of the series resistance measured is of interest and is shown in Fig. 5 for a thallium-doped sample at zero d.c. bias. It can be seen that heating from 25°C to 122°C only causes a 10 per cent reduction in the series resistance. This resistance is determined by the resistivity of the selenium (assuming that an ohmic contact is made by the nickel). The constancy of the resistance as a function of voltage that was shown in Fig. 2 confirms that the nickel was making an ohmic contact. The constancy of the resistance measured agrees with results of previous workers[2, 3] in that the resistivity of hexagonal selenium is only slightly dependent on temperature. The interesting point here is that this result has been obtained with a sample having one blocking and one injecting electrode. The agreement of the results with those of previous workers adds strength to the series resistance/capacitance model of the sample shown in Fig. 1. The transient response of the system is interpreted in terms of this model.

TRANSIENT MEASUREMENTS

Figure 6 is a block diagram of the circuit used to measure the transient response of the sample to illumination. A 50-MΩ resistance in series with the 1 kHz source converted it into a constant current source so that any change

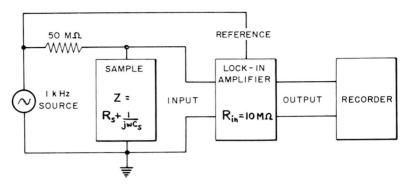

Fig. 6. Circuit for measurement of transient capacitances.

in the sample impedance would show up as a voltage change across the sample. A typical value for the sample impedance was 1 MΩ so that the 10-MΩ input impedance of the lock-in amplifier produced negligible additional loading on the source. The amplifier was tuned to measure the component of impedance 90° out of phase with the applied voltage so that a direct measure of $1/\omega C_s$ was obtained. Figure 7 is a typical response taken when a

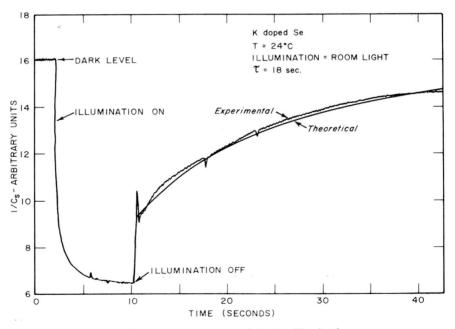

Fig. 7. Decay of capacitance following illumination.

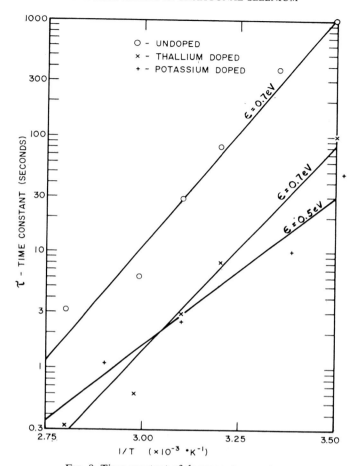

Fig. 8. Time constant of decay vs. temperature.

potassium-doped sample was illuminated with white light under zero bias conditions.

Upon illumination of the barrier region, the effective density of acceptors is increased and the capacitance increases to a new non-thermal equilibrium value.[4] When the illumination is removed, the non-thermal carrier concentration decays exponentially back to zero with a time constant determined by the thermal rate of release from the deep acceptors. Consequently

$$\Delta\left(\frac{1}{C_s^2}\right) = \Delta\left(\frac{1}{C_s^2}\right)_0 \exp^{-t/\tau}$$

To show the degree of fit, a calculated decay curve corresponding to a time constant at 18 seconds has been included with the actual recorder trace in Fig. 7. In practice the 50 per cent change in $1/C_s$ has been used to define the time constants shown in Fig. 8. As can be seen from Fig. 8, time constants from 0.3 to 1000 seconds have been measured using this technique on a number of samples. In contrast to the measurements described earlier, the rate of decay is markedly dependent on temperature defining an energy

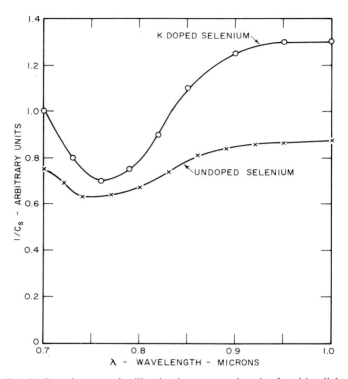

FIG. 9. Capacitance under illumination vs. wavelength of exciting light.

level 0.7 eV above the valence band for pure and thallium-doped samples and 0.5 eV for the potassium-doped samples. The rate of decay is ten times as fast for the thallium-doped samples as for pure selenium.

Measurements have also been made of the capacitance response to monochromatic excitation to estimate the energy difference between the deep levels and the conduction band. As can be seen from Fig. 9, both pure and potassium-doped samples showed a cut-off between 0.9 and 1 micron. Adding the thermal activation and optical excitation energies leads to an estimated bandgap of between 1.86 and 1.98 eV for hexagonal selenium.

REFERENCES

1. HENISCH, H. K., *Rectifying Semiconductor Contacts*, O.U.P., Oxford, 1957, p. 214.
2. HENKELS, H. W., *J. Appl. Phys.* **22**, 916 (1951).
3. PLESSNER, K. W., *Proc. Phys. Soc. (London)*, **B64**, 681 (1951).
4. WILLIAMS, R., *Bull. Am. Phys. Soc. II*, **11**, 754 (1966).

DISCUSSION

SIEMSEN: What was the bulk resistivity of the selenium single crystals that you used?

LANYON: The resistivities agreed well with typical values for selenium single crystals, namely of the order 10^5 to 10^6 ohm-cm. Where the resistivity was also checked with a four-point probe, the agreement between the two estimates was very good.

CHAMPNESS: How do you explain the slope change of $1/C_s^2$ vs. V with distance from the surface? Do you think that diffusion of the metal into the selenium occurs?

LANYON: The slope of the $1/C_s^2$ vs. V curve gives a measure of the doping content as a function of position. In Figs. 2 and 4 I merely drew two tangents to estimate typical values of densities. The density was always greater close to the surface than in the bulk. There is no reason to ascribe this change in density to diffusion of the metal although this could occur in principle. It would seem that surface damage is an equally likely cause, particularly since this number occurs with all types of samples.

BECKER: Most people who have measured the a.c. resistance of trigonal selenium crystals have observed a strong frequency dependence (usually interpreted as barriers or trapping). Did you observe a frequency dependence for R_s and how did you interpret it?

LANYON: We did not observe any significant variation of R_s in the frequency range from 1 Hz–1 kHz. However, we did find that in some frequency ranges samples showed a dissipation factor *independent* of frequency. It is too early to comment on the significance of this observation at the present time.

STUKE: Does your decay of the carrier concentration after illumination follow a relation which can be explained by a single time constant?

LANYON: We did not check on the time relationship on every occasion, normally taking the time for $\Delta(1/C_s)$ to decay to one-half of its initial value as the time constant. On each occasion that we fitted to the measured decay curve we obtained a fit equivalent to that shown in Fig. 7. It must be remembered that one is looking at a different effect from the one that is measured in the decay of photocurrent. In this latter measurement all carriers are weighted equally. In the experiment described here each carrier is weighted inversely according to the distance from the aluminum electrode (anode). Consequently the motion of carriers during their first release time dominates the decay of the signal. At long times the fit between the experimental and theoretical curves is lost in noise.

KOLB: Isn't it possible that there is still surface damage remaining in both the potassium- and thallium-doped samples which makes them appear to be the same? They may indeed still both be damaged to the same depth.

LANYON: That is true.

MORT: Are the acceptor concentrations which you observed consistent with the impurity concentrations of the crystals used? Can you conclude from your measurements that thallium and potassium introduce acceptor levels into trigonal selenium?

LANYON: The acceptor concentrations are approximately four orders of magnitude smaller than the doping concentrations so that most of the dopant is inactive. From our measurements it seems that the thallium increases the density of levels 0.7 eV above the valence band by a factor of about 10. The potassium introduces a new level of 0.5 eV. The effects seem to be surprisingly small.

LAUDISE: Could you mention any other differences between Keezer's crystals and Harrison's crystals other than those that you have mentioned in the paper? Keezer's crystals contain thallium or potassium and Harrison's crystals more lineage.

LANYON: We have found no other definite correlations at the present time. From discussion it appears that there are quite significant differences between the mechanical properties. Harrison's crystals are brittle whereas Keezer's crystals deform easily. Harrison suggests that this could be due to decoration of the dislocations by the impurity. Certainly little of the thallium is electrically active.

KEEZER: Potassium-doped samples are more perfect than thallium-doped samples.

Na_2S etches selenium on $(10\bar{1}0)$ face at about 100–200 μ/min. H_2SO_4 etches tellurium on $(10\bar{1}0)$ face at about 1–5 μ/min.

PSEUDOPOTENTIAL BAND STRUCTURE FOR SELENIUM

ROLF SANDROCK

The James Franck Institute, University of Chicago, Chicago, Illinois 60637

INTRODUCTION

In recent years, a number of elaborate calculations[1-3] have been published on the electronic structure of hexagonal selenium. While these were able to explain some of the experimental information accumulated at the same time, they depend on rather restrictive approximations like the assumption of a muffin-tin potential and the neglect or very crude treatment of exchange interaction.

The pseudopotential approach,[4] less ambitious than the "first-principles" calculations, is based on empirical form factors and thereby avoids many of the approximations. Therefore, it could serve as a bridge between the theoretical models and experiment. Thus the purpose of this paper is twofold: to check the validity of previous theoretical results and to contribute some new information that might be useful to interpret optical measurements.

THE PSEUDOPOTENTIAL

From experimental data on optical transitions, Cohen and Bergstresser[5] have derived and compiled pseudopotential form factors for a number of semiconductors, both diamond-type elements and zincblende-type compounds.

This information can now be exploited to calculate energy bands for other crystals involving the same constituents.[6] Basically, no further adjustment should be necessary with this approach.

Thus the atomic form factors $v_{Se}(q)$ were obtained from

$$v_{Se}(q) = C(v^s(q) - v^a(q)),$$

where v^s and v^a are the symmetric and antisymmetric parts, respectively, of Cohen and Bergstresser's[5] form factors for ZnSe. $C = \frac{3}{2}\Omega_{ZnSe}/\Omega_{Se}$ corrects for the different unit cell volumes and numbers of atoms per unit cell in the selenium and zinc selenide structures. Multiplying $v_{Se}(q)$ by the appropriate structure factor

$$s_{Se}(q) = \tfrac{1}{3} \sum_{i=1}^{3} e^{iq \cdot t_i}$$

(where t_i denotes the location of the ith atom in the cell), we get the pseudopotential coefficients for hexagonal selenium.

Of course, this procedure requires interpolation between Cohen and Bergstresser's $v(q)$ and even extrapolation beyond their q range. The first reciprocal lattice vectors for selenium are shorter than the (111) vector for ZnSe. But this extrapolation ambiguity does not affect the more relevant bands to within 0.01 ryd. When extrapolating $v(q)$ towards large q, the model potential of Animalu and Heine[7, 8] was followed as a guideline (Fig. 1).

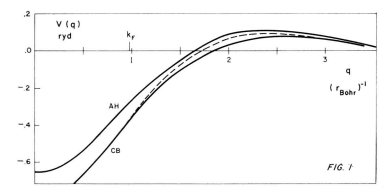

FIG. 1. Pseudopotential form factors for selenium: AH—model potential after Animalu and Heine;[7, 8] CB—derived from Cohen and Bergstresser;[5] dashed line—adjusted pseudopotential.

Using the pseudopotential just described, without any adhoc adjustment, the band structure shown in Fig. 2 and discussed in more detail below was obtained. It shows a direct energy gap near H of $E_G = 1.4$ eV. This is in fair agreement with the experimental value $E_G = 2.0$ eV.[9] After all, we have made a drastic change in crystal structure when going from ZnSe to Se, much more drastic than going from zincblende to wurtzite structures.[6] We have put the selenium atom in a completely different environment, certainly effecting a considerable charge transfer.

However, these differences can be accounted for by a moderate adjustment of the form factors (Fig. 1). Since there are so many different reciprocal lattice vectors involved, there is no point in fitting just one or a few of them to obtain the experimental gap. Instead, this was achieved by slightly raising the entire form factor curve for $q > k_F \approx 1 r_B^{-1}$. (Note that the Cohen and Bergstresser form factors are only accurate to within 0.01 ryd anyway.) The adjusted pseudopotential lies between the ones of Cohen and Bergstresser[5] and Animalu and Heine,[7, 8] indicating that the dielectric screening seems to be somewhat more metallic in elemental selenium than it is in zinc selenide.

The pseudopotential employed in either form is a local one. It was felt that

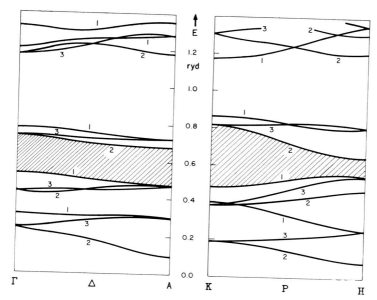

FIG. 2. Energy bands for selenium using the pseudopotential derived from Cohen and Bergstresser.[5] The shaded region separates valence and conduction bands, the numbers 1, 2, 3 denote point group representations for the axes Δ and P.

non-locality should have a negligible effect in the present case where both valence and conduction states have predominantly p-character. Also, spin-orbit interaction has not been taken into account. Splittings due to spin are expected to be small for selenium, though they may be important for tellurium.

CALCULATION AND RESULTS

Structure and pseudopotential of selenium require that a larger set of plane waves be included in the secular equation than had been necessary for cubic semiconductors.[10] Recourse had to be made to the threefold symmetry, therefore only the axes Δ and P are shown in Figs. 2 and 3. Along these axes, about forty symmetrized plane waves were taken into account, while sixty more of them were included approximately by Löwdin perturbation (see, e.g., Brust[10]). These numbers correspond to cut-off energies of 5.5 ryd and 10 ryd, respectively. At this stage, all the relevant energy band separations show convergence to within 0.1 eV.

Except for the fitting of the gap near H, our band model is basically the same for either choice of the pseudopotential (Figs. 2 and 3). However, the gap has almost become indirect in the adjusted case. The model is also similar to the one obtained earlier by the no-spin KKR calculation,[3] with two notable differences:

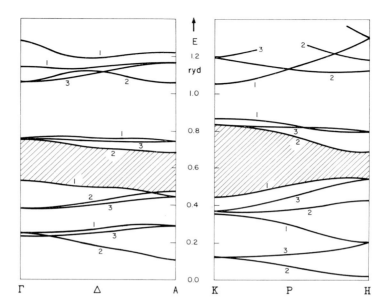

FIG. 3. Energy bands for selenium using the adjusted pseudopotential.

(1) All bandwidths and interband separations have nearly doubled their values on the energy scale. This confirms the observation (made, e.g. by Rössler and Lietz[11]) that muffin-tin potentials, although otherwise capable of good results,[12] sometimes tend to make bands too flat. This is especially true when the muffin-tin spheres fill only a small fraction of the cell (24 per cent for selenium). Our result also justifies the approach of Mohler, Stuke and Zimmerer[13] who interpreted their reflectivity measurements in terms of Treusch and Sandrock's band model[3] assuming an overall scaling factor to obtain quantitative agreement. As a consequence of the increased bandwidths, the upper and lower valence band triplets overlap near K. According to Mohler et al.[13] these two triplets should be responsible for the reflectance spectrum falling into two distinct structures below and above 6.5 eV. Since there is probably a gap between the upper and lower valence bands over large parts of the Brillouin zone, the present model may give support to this explanation. However, it is obvious from Figs. 2 and 3 that higher conduction bands should also contribute to the ultraviolet reflectivity above 8 eV. Finally, one would expect that broader bands have smaller effective masses. In fact, it is found for the top of the valence band, $m_{\parallel} = 0.4\, m_0$ as compared to $m_{\parallel} = 1\, m_0$ in the earlier model.[3]

(2) The other striking disagreement with the earlier model[3] is found at point H. The KKR results suggest that the perpendicular-allowed absorption

edge $H_3 \rightarrow H_1$ should be followed by another direct transition $H_2 \rightarrow H_1$ which is allowed for parallel polarization and only 0.1 eV higher in energy. The pseudopotential results, however, show that the H_2 level is about 1 eV below the H_1 edge. Therefore indirect absorption will definitely start at energies much smaller than the one required for the $H_2 \rightarrow H_1$ transition. This agrees with recent measurements by Roberts et al.[14] It is still believed that the perpendicular absorption edge is due to direct transitions near H, though the adjusted model (Fig. 3) reveals that there is an indirect gap along the Δ-axis which is virtually degenerate with the direct one. This situation might be responsible for the permanent controversy concerning direct vs. indirect absorption.

There is little doubt that the majority of the reflectivity peaks observed[9, 13] will be contributed to by large parts of the Brillouin zone, not just by a few symmetry points or axes. Transitions at general points are not likely to show pronounced selection rules, though the oscillator strengths will of course be different for different polarization. In fact, the reflectivity spectrum[13] in the region between 3 eV and 5 eV looks quite similar for $E \| c$ and $E \perp c$, though different in height. Hence full understanding of the optical spectrum will require knowledge of the bands at a large number of points in k-space.

ACKNOWLEDGMENTS

This research was supported by the Army Research Office (Durham). The author also benefited from the use of facilities provided by the Advanced Research Projects Agency for materials research at the University of Chicago.

Gratitude is expressed to Professor J. C. Phillips and Dr. Fred Mueller for many fruitful discussions and suggestions.

REFERENCES

1. OLECHNA, D. J. and KNOX, R. S., *Phys. Rev.* **140**, A986 (1965).
2. HULIN, M., *J. Phys. Chem. Solids* **27**, 441 (1966).
3. TREUSCH, J. and SANDROCK, R., *Phys. Status Solidi* **16**, 487 (1966).
4. PHILLIPS, J. C. and KLEINMAN, L., *Phys. Rev.* **116**, 287 (1959).
5. COHEN, M. L. and BERGSTRESSER, T. K., *Phys. Rev.* **141**, 789 (1966).
6. BERGSTRESSER, T. K. and COHEN, M. L., *Phys. Letters* **23**, 8 (1966).
7. ANIMALU, A. O. E. and HEINE, V., *Phil. Mag.* **12**, 1249 (1965).
8. ANIMALU, A. O. E., Techn. Report No. 4, 1965, Solid State Group, Cavendish Laboratory, Cambridge, England.
9. TUTIHASI, S. and CHEN, I., *Phys. Rev.* **158**, 623 (1967).
10. BRUST, D., *Phys. Rev.* **134**, 1337 (1964).
11. RÖSSLER, U. and LIETZ, M., *Phys. Status Solidi* **17**, 597 (1966).
12. ECKELT, P., MADELUNG, O. and TREUSCH, J., *Phys. Rev. Letters* **18**, 656 (1967).
13. MOHLER, E., STUKE, J. and ZIMMERER, G., *Phys. Status Solidi* **22**, K49 (1967).
14. ROBERTS, G. G., TUTIHASI, S. and KEEZER, R. C., *Solid State Comm.* **5**, 517 (1967).

DISCUSSION

BECKER: Can you state what changes would have to be made in your calculations to obtain an indirect edge?

SANDROCK: In my results, lowering the conduction band at A by about 0.05 eV would be sufficient to make the absorption edge indirect. This is less than the accuracy that is claimed. Therefore, it cannot be decided definitely whether or not the edge is direct for perpendicular polarization. For parallel polarization it clearly seems to be indirect.

SIEMSEN: We have recently been doing some absorption measurements with trigonal selenium crystals. In these experiments the light beam is parallel to the c-axis, the **E** vector being fixed in the plane perpendicular to the axis. If we now turn the crystal around the c-axis it seems that we find a 60° periodicity instead of the expected planar isotropy. Could your band model provide an explanation for this result? Our measurements are only preliminary, though, and the anisotropy might be due to other effects.

SANDROCK: The present band calculation, as well as previous ones, does not include enough points to answer this question. However, the dielectric tensor has to be isotropic in the (0001) plane, since selenium is a uniaxial crystal. Hence the absorptivity should also be rotational-isotropic.

LUCOVSKY: The observed effect might be due to the dichroism of selenium.

OPTICAL PROPERTIES AND ENERGY BAND STRUCTURE OF TRIGONAL SELENIUM

W. HENRION

Physikalisch-Technisches Institut, Deutsche Akademie der Wissenschaften zu Berlin, Berlin—DDR

INTRODUCTION

Considerable work has been done in recent years on the calculation of the band structure of trigonal selenium. The present situation is, in spite of these attempts, rather unsatisfactory. Therefore, optical investigations on single crystals are of special significance, because the energetic position of critical points may be determined experimentally with their help. The optical investigations presented were performed on single crystals grown from the vapor phase by sublimation.[1, 2] These crystals have very good natural surfaces (prism plane ($10\bar{1}0$)). A mechanical or chemical treatment of the crystals was not necessary.

INVESTIGATION OF THE ABSORPTION EDGE

The behavior of the absorption at the fundamental lattice edge of selenium has been explained repeatedly by non-vertical transitions.[3–5] Measurements performed with polarized light in the temperature region of 400° to 90°K gave no indication of a non-vertical indirect absorption.[6] If such an absorption mechanism existed, we should expect a plot of $\alpha^{\frac{1}{2}}$ vs. the photon energy $h\nu$ to be built up of linear pieces having more or less pronounced break points at energies corresponding to the threshold values of the respective processes. Such a fine structure could not be observed. The behavior at lower temperatures (the curves become steeper) also supports the suggestion that the absorption edge of selenium is based on vertical transitions. Figure 1 illustrates that in the temperature region investigated, the absorption curve increases exponentially within a photon energy interval of more than 0.1eV independent of the polarization direction. The absorption at energies lower than the edges is between 2 and 20 cm^{-1} remaining almost constant to 20 microns. The exponential absorption edge is relatively well described by the half-empirical formula of Urbach,[7]

$$\alpha = \alpha_0[1 + \exp \beta(E_g^* - h\nu)]^{-1}$$

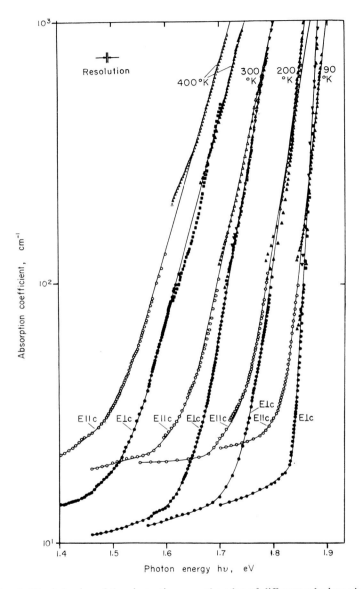

FIG. 1. The behavior of the absorption near the edge of different selenium single crystals for $E \parallel c$ and $E \perp c$ (multiple reflections were taken into account in the calculation of α).

where $\beta = \sigma/kT$ for temperatures not too low; α_o, σ, E_g^* are constants characteristic of the crystal. E_g^* is approximately equal to the width of the forbidden band E_g. The slope β for $E \perp c$ increases from 17.5 eV^{-1} at 400°K to 85.3 eV^{-1} at 90°K, i.e. $\sigma \approx 2/3$; for $E \| c$ it rises from 16.7 eV^{-1} at 400°K to 45 eV^{-1} at 90°K, i.e. $\sigma \approx 1/2$. A calculation of E_g^* assuming $\alpha_o = 10^6$ cm^{-1} for $E \| c$ and $\alpha_o = 2 \times 10^5$ cm^{-1} for $E \perp c$[8] yields at 300°K $E_{g\,\|}^* = 2.11$ eV and $E_{g\,\perp}^* = 1.98$ eV in very good agreement with the reflectivity measurements described below. The exciton resonance observed there already at 300°K may cause the exponential edge by an exciton-lattice interaction. In covalent semiconductors with a direct absorption edge, however, it is more probable that the exponential tail absorption is due to charged impurities (internal Franz–Keldysh effect).[9] Roberts, Tutihasi and Keezer[10] recently confirmed results obtained in this laboratory in the case $E \perp c$. For $E \| c$ they attributed the absorption to phonon-assisted transitions. However, this opinion cannot be shared, not only for the reasons mentioned above, but also because in the case $E \| c$, too, a strong maximum in reflectivity is found. It is possible that this question may be solved unequivocally by thermal-absorption measurements.[11]

REFLECTIVITY IN THE INTERBAND REGION

The reflectivity measurements in this study were carried out with polarized light at almost perpendicular incidence in the fundamental absorption range. In Figs. 2–4 the experimental data for 300°K and 90°K are plotted. All measuring points are obtained by taking the average of two measurements. For $E \| c$ the reflectivity is throughout higher than for $E \perp c$, and an intersection of both curves does not occur. The first maximum for $E \| c$ at 300°K is located at about 2.1 eV. On cooling down it shifts about 6×10^{-4} eV/deg towards higher energies. The first maximum for $E \perp c$ at 300°K occurs at 1.95 eV (Fig. 2). At this polarization also a sub-maximum at 1.98 eV (see Fig. 2) and a shoulder at about 2.1 eV are observed in contrast to other authors.[12–14] The latter occurs at the same energy as the maximum for $E \| c$. On lowering the temperature 210°K a distinct peak is observed at 1.95 eV, whereas the peak at 1.98 eV is no longer recognizable. Upon further cooling the maximum increases and towards higher energies the formation of an anti-resonance is observed. The shape of the maximum, especially its strong temperature-dependence and the even stronger temperature-dependence of the anti-resonance are characteristic of an exciton. Evidently, this exciton may be recognized already at 300°K in spite of strong broadening. The sub-maximum at 1.98 eV then ought to be due to the interband transition itself, because the higher excitation levels of the exciton are certainly not observable. On the other hand, a weak structure found at 90°K and 2.01 eV may be due to a higher excitation level of the exciton ($n = 2$) or the interband transition. The

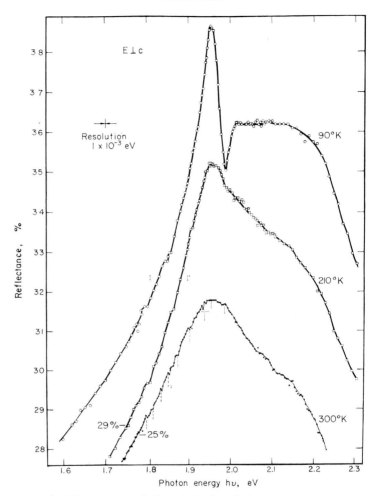

FIG. 2. Reflectivity spectrum of trigonal selenium single crystals at the fundamental absorption edge for $E \perp c$ (relative measuring accuracy is shown beside the curves).

temperature coefficient of the first interband transition for $E \perp c$ would then amount to about 1.5×10^{-4} eV/deg being clearly smaller than that for $E \| c$.

In the present situation a correspondence of those primary structures in the vicinity of the absorption edge to interband transitions in one of the band models calculated for selenium till now is difficult to establish. The model calculated with the Kohn–Rostoker method by Treusch and Sandrock[15] has proved to be the most appropriate (Fig. 5). In the last column of Table 1 the strongest maxima with transitions in this model are identified. The establishment of this correspondence is suggested by calculated selection rules and by the observed polarization and temperature dependence of the maxima. However, this correspondence should not be regarded as certain, because if spin–orbit interaction is accounted for in the theory, this may bring about essential

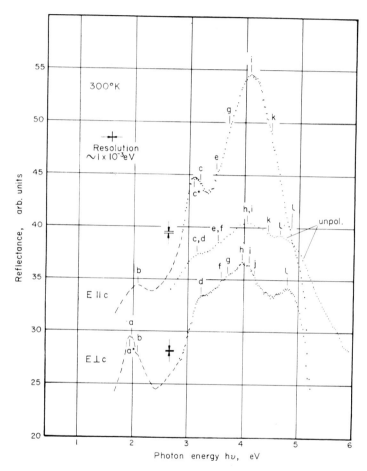

FIG. 3. Reflectivity spectrum of trigonal selenium single crystals at 300 K for $E\|c$ and $E\perp c$.

modifications of the band model. According to Fig. 5 the smallest energy gap is located in point H of the Brillouin zone or on the P-axis in the direct environment of point H. In point H the interband transitions H_3–H_1 ($E\perp c$) and H_2–H_1 ($E\|c$) with $E_{g\perp} < E_{g\|}$ are possible. $E_{g\perp} < E_{g\|}$ holds likewise, if the first transition for $E\perp c$ is directed from P_1 to P_2. Towards higher energies first a peak for

$E\|c$ at 3.1 eV (Fig. 4) is observed, probably caused by a saddle-point exciton. The corresponding interband transition at the peak c is located at 3.2 eV (Z_2–Z_1) and does not shift essentially with altering temperature. At 300°K c is

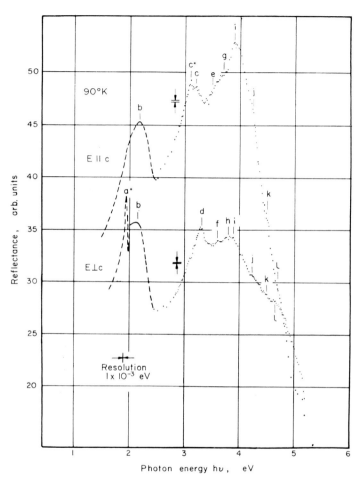

FIG. 4. Reflectivity spectrum of trigonal selenium single crystals at 90°K for $E\|c$ and $E\perp c$.

only recognizable as a shoulder of c^* (Fig. 3). The next pronounced maximum at 4.1 eV is obtained for both polarization directions, and upon cooling it also shifts for both polarizations by about 1×10^{-3} eV/deg towards lower energies. We suppose this maximum to be caused by the same transition Γ_3–Γ_3 which is allowed for both polarizations.

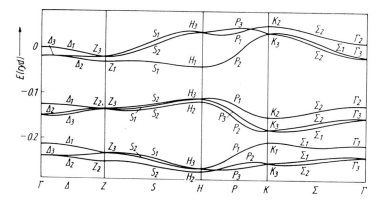

FIG. 5. Band structure of trigonal selenium according to Treusch and Sandrock.[15]

TABLE 1. *Energies of Observed Reflectivity Maxima and Their Polarizations*

Maximum	Energy (eV) 300°K	90°K	Observed for polarization	Possible transition (ref. 15)
a^*	1.95	1.953	$\perp c$	Exciton
a	1.98	2.01	$\perp c$	H_3-H_1
b	2.11	2.22	$\perp, \parallel c$	H_2-H_1
c^*	3.09	3.10	$\parallel c$	Exciton
c	3.20	3.20	$\parallel c$	Z_2-Z_1
d	3.24	3.29	$\perp c$	Z_3-Z_1
e	3.5	3.5	$\parallel c$	
f	3.6	3.6	$\perp c$	$\Gamma_2-\Gamma_3$
g	3.7	3.7	$(\perp), \parallel c$	
h	3.98	3.80	$\perp c$	
i	4.10	3.90	$\perp, \parallel c$	$\Gamma_3-\Gamma_3$
j		4.2	$\perp, \parallel c$	
k	4.5	4.5	$(\perp), \parallel c$	
l	4.8	4.7	$\perp, (\parallel)c$	H_2-H_3

Apart from the reflectivity peaks mentioned in Table 1 some weaker structures were also observed. Thus, the reflectivity for $E \perp c$ exhibits a fine structure between 1.8 and 1.95 eV (Fig. 2). Other structures are found between 2.4 and 3.1 eV. In this region according to Chen[16] the electroreflectance of polycrystalline selenium layers also exhibits weak peaks at almost the same energies at which they have been observed in this laboratory. Unfortunately such investigations carried out with polarized light have not been reported up to the present, because selenium single crystals are obviously too high-ohmic to allow the application of the electroreflectivity method. The thermoreflectivity measurements carried out till now[17] solely proved the applicability of this new differential method to high-ohmic semiconductors like selenium.

INTRABAND TRANSITIONS IN SELENIUM

If the anisotropy at the absorption edge with $E_{g\perp} < E_{g\parallel}$ is caused by valence band splitting, one may expect transitions between those sub-bands of the valence band in p-type selenium. Some of the absorption bands in the near infrared region observed on selenium single crystals have been explained in recent years in such a manner.[4, 5, 18] What is observed here, are two double bands at 2.9/3.4 microns. Whereas Gobrecht and Tausend[4] observed a polarization dependence of the bands, work in this laboratory indicated a

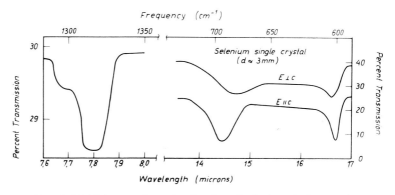

FIG. 6. Multiphonon absorption in trigonal selenium single crystals.

possible connection between band intensity and the chlorine contents of the initial material.[5] In photoabsorption, Kessler and Sutter[18] observed these bands and two additional ones at 1.9 and 4.05 microns. These authors also suppose these bands to have appeared in the photofluorescence.[19] It seems to be established now, however, that three bands are due to contaminants absorbed by the crystal surface.[20] The bands at 1.9, 2.9 and 3.4 microns are also obtained, if the selenium crystal in the optical path of the spectrograph is replaced by any other infrared transmitting substance (e.g. plates of KBr or NaCl).† How these bands appear also in photoabsorption has not yet been fully explained. It is possible, on the other hand, that they are also observed in reflectivity measurements.[4]

† The 2.9-micron band is undoubtedly identical with the OH-stretching band slightly modified by hydrogen-bonding.[21] The 1.9 micron band is far less intensive. It was also found upon examination of capillary water layers.[22] The 3.4-micron band is probably due to a C–H vibration arising from hydrocarbons absorbed from the atmosphere.[23]

Some weak absorption bands have been found in the infrared region which undoubtedly may be attributed to selenium (Fig. 6). They depend only weakly on polarization. Whereas the 14.4-micron and 16.7-micron bands are higher in the case of $E\|c$, the 7.8-micron band is higher for $E\perp c$. The absorption coefficient alters by about 3 cm^{-1} at the 14.4/16.7-micron bands and at the 7.8-micron band only by 0.1 cm^{-1} at a background of 3 cm^{-1}. Whereas the band complex at 14.4/16.7 microns may be considered to arise from three-phonon processes,[24, 25] the 7.8-micron band probably represents a harmonic of this process (six phonons!). An absorption arising from intraband transitions between sub-bands of the valence band has so far not been observed.

REFERENCES

1. ECKART, F., *Recent Advances in Selenium Physics*, Pergamon, London, 1965.
2. ECKART, F., HENRION, W. and PEIBST, H., *Z. Phys. Chem.* **227**, 93 (1964).
3. CHOYKE, W. J. and PATRICK, L., *Phys. Rev.* **108**, 25 (1957).
4. GOBRECHT, H. and TAUSEND, A., *Z. Physik* **161**, 205 (1961).
5. ECKART, F. and HENRION, W., *Phys. Status Solidi* **2**, 841 (1962).
6. HENRION, W., *Phys. Status Solidi* **12**, K113 (1965).
7. MOSS, T. S., *Optical Properties of Semiconductors*, Butterworths, London, 1959.
8. PROSSER, V., *Czech. Z. Physik* **B10**, 306 (1960).
9. REDFIELD, D. and AFROMOWITZ, M. A., *Appl. Phys. Letters* **11**, 138 (1967).
10. ROBERTS, G. G., TUTIHASI, S. and KEEZER, R. C., *Solid State Comm.* **5**, 517 (1967).
11. BERGLUND, C. N., *J. Appl. Phys.* **37**, 3019 (1966).
12. STUKE, J. and KELLER, H., *Phys. Status Solidi* **7**, 189 (1964).
13. TUTIHASI, S. and CHEN, I., *Phys. Rev.* **158**, 623 (1967).
14. MOHLER, E., STUKE, J. and ZIMMERER, G., *Phys. Status Solidi* **22**, K49 (1967).
15. TREUSCH, J. and SANDROCK, R., *Phys. Status Solidi* **16**, 487 (1966).
16. CHEN, J. H., *Phys. Letters* **23**, 516 (1966).
17. LANGE, H. and HENRION, W., *Phys. Status Solidi* **23**, K67 (1967).
18. KESSLER, F. R. and SUTTER, E., *Z. Physik* **173**, 54 (1963).
19. KESSLER, F. R. and SUTTER, E., *Phys. Status Solidi* **23**, K25 (1967).
20. HENRION, W., *Phys. Status Solidi* **14**, K51 (1966).
21. MURR, A., Dissertation, München University, 1964.
22. CURCIO, I. A. and PETTY, C. C., *J. Opt. Soc. Am.* **41**, 302 (1951).
23. HARRICK, N. J., *Appl. Optics* **4**, 1664 (1965).
24. LUCOVSKY, G., MOORADIAN, A., TAYLOR, W., WRIGHT, G. B. and KEEZER, R. C., *Solid State Comm.* **5**, 113 (1967).
25. GEICK, R., SCHRÖDER, U. and STUKE, J., *Phys. Status Solidi* **24**, 99 (1967).

CRYSTAL GROWTH
AND CHARACTERIZATION

IMPURITIES IN SELENIUM—DETECTION AND INFLUENCE ON PHYSICAL PROPERTIES

H. GOBRECHT

II. *Physikalisches Institut der Technischen Universität Berlin, Berlin, Germany*

IT IS well known that small amounts of chlorine raise the electrical conductivity of selenium by two orders of magnitude. Stoichiometric amounts of sodium, thallium, or mercury cancel this effect. Small amounts of iron, however, increase it. Since these and similar effects are not understood, it is important to study the role of impurities in selenium. First of all one must know the kind and the amounts of trace impurities in selenium.

One of the most sensitive trace analysis methods is neutron activation analysis. In the last three years, new methods have been developed in this laboratory for the determination of chlorine, antimony and sulfur in selenium with detection limits of 0.01, 0.001 and 0.05 ppm respectively.[1, 2] Efforts have been made to determine additional elements, but a consideration of the preconditions which make neutron activation analysis of selenium possible showed that it is not applicable to all elements. Neutron irradiation induces a high activity in selenium itself. Apart from Se-75, the half-lives of all nuclides extend from a few seconds to 3 hr. Therefore, the determination of elements which do not yield radionuclides with half-lives longer than 3 hr is difficult, unless there is a particular chemical behavior, as, for example, in the case of chlorine, which allows a rapid separation. Generally speaking, the development of the "normal" activation analysis for selenium, i.e. analysis by (n,γ)-activation, has ceased, because those elements that are of any importance for selenium and fulfil the preconditions have been determined. There are some important elements, however, especially nonmetallic elements, the determination of which is prevented by the rapid decay of their radionuclides. Unfortunately, the determination of traces of nonmetals by conventional methods, e.g. by spectral analysis, is difficult, too.

The difficulties in activation analysis will be explained by consideration of the determination of oxygen. Oxygen consists of three isotopes O-16, O-17, O-18. The only (n,γ)-process which leads to a radionuclide is the reaction O-18(n,γ)O-19. The saturation activity is very low, however, because of the O-18 abundance of 0.2 per cent and its cross-section for thermal neutrons of

0.21 mbar. The half-life is very short, too (29.5 sec), and altogether it is impossible to determine oxygen by (n,γ)-processes.

A different approach is provided by reactions with fast neutrons, e.g. (n,p)-reactions. On account of its abundance, only O-16 would give a sufficiently high activity via O-16(n,p)N-16. But the half-life of N-16 of 7.4 sec again forbids its application in activation analysis, unless one uses a neutron generator for fast neutrons. There is a certain possibility of determining oxygen traces, because the time between activation and measurement in a γ-spectrometer can be made very short.

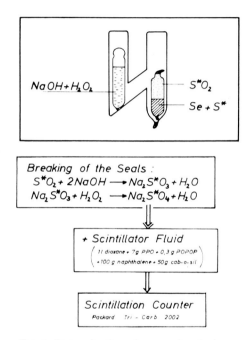

FIG. 1. Determination of oxygen in selenium.

In order to avoid the difficulties mentioned above, a new method was developed for determining oxygen in selenium which is based on a conventional chemical analysis. Advantage is taken of the radiochemistry and the oxygen atoms are labelled via radioactive sulfur-35 through the formation of gaseous sulfur dioxide, which can be measured in a scintillation counter. For this purpose selenium and excess sulfur are heated in a baked-out ampoule. It was ascertained in a preliminary test by infrared absorption measurements that the oxygen that is present in the form of selenium dioxide is attracted by sulfur forming SO_2.

The operating sequence of this experiment, which is still in the stage of development, is illustrated in Fig. 1. In the right leg of a U-shaped reaction container there is an ampoule containing the selenium and the SO_2; in the left one there is a mixture of sodium hydroxide and hydrogen peroxide serving as absorbing fluid for SO_2. By turning the container the breaking seals of the ampoules are opened and the gas can react with the solution forming sodium sulfate. The absorption time is about 20 hr. After that the radioactive solution is added to a scintillator fluid sensitive to the β-radiation of S-35.

The activity as an equivalent to the oxygen content is measured in a liquid scintillation counter (Packard Tri-Carb 2002). Originally, the activity of the radioactive gas was measured in an ionization chamber. But as the SO_2 contaminated the chamber more and more, recourse was made to the more reproducible method of scintillation counting. The total procedure was calibrated by samples with known oxygen concentration. At the moment the limit of detection is 1 ppm. This limit is not caused by a low activity (more than 10,000 pulses per minute were obtained), but by the deviations of the results. The main reasons for this fact are as follows:

1. Although the influence of adsorbed oxygen is minimized by the tracer method, there still remains oxygen adsorbed on the surfaces of the selenium pellets, the added sulfur, and the wall of the ampoule.
2. There are other elements, for example hydrogen, which form gaseous compounds with sulfur. Thus, a blank value is measured which permits a correction for adsorbed oxygen and other elements such as hydrogen which may interfere with the determination.

If better control of the disturbing influences is realized, the accuracy and with that the detection limit can be much improved. An attempt is being made to find similar methods for other important elements. Oxygen and chlorine can also be found by heating the selenium under a vacuum. These impurities are pumped off, compressed and can be seen in a spark discharge. This method is sensitive, too, but in a mass spectrometer it was noted that not every molecule of the various selenium compounds, e.g. Se_2Cl_2, is decomposed by annealing. The chlorine content could be reduced by distillation, e.g. from 0.8 ppm to 0.26 ppm.

EPR measurements are very sensitive, too. Of course, they are only applicable if the trace elements possess an unpaired electron. Very strong, but broad signals were obtained for the elements with incomplete $3d$ shells as Fe, Ni, Cr, Mn ($g = 2.27$). Unfortunately no separation could be made amongst these elements. It can only be said that metals of the transition elements are present as impurities. In pure amorphous selenium a sharp signal ($g = 2.0038$) was obtained.

It has been assumed that SeO_2^- is responsible for this sharp peak.[3-5] But a distillation under vacuum could not reduce this peak.

The knowledge of the phenomena concerning self-diffusion and diffusion of impurities is of great importance, too (e.g. for rectifiers).

Although there are some diffusion investigations on polycrystalline selenium, no definite relations to the known properties of selenium can be stated. The measured diffusion coefficients turned out to be very structure-sensitive. Particularly accurate information on well-grown single crystals is missing. The analysis of diffusion experiments with polycrystals is complicated, because there is an interference of volume diffusion, grain boundary diffusion, and diffusion along inner surfaces such as pores and cracks.

The samples used in the present study were produced in the following manner: high purity selenium was vacuum-distilled and transformed to the trigonal modification by a two-stage annealing process (15 min at 110°C and 60–70 hr at 180–200°C, at a pressure of about 10^{-5} torr). The polycrystalline material was powdered and by passing through sieves divided into fractions of definite size. Cylindrical pellets were pressed under vacuum by a pressure of 12 kbar (diameter 10 mm, thickness 3–5 mm). They were annealed once more at 190°C for 100 hr. The perfect crystallinity was confirmed by X-ray investigations. Measurements of the density yielded 93–96 per cent of the theoretical density, i.e. only a little porosity was present in the samples. Thus an influence of diffusion processes along inner surfaces was largely excluded.

The radiation properties of the radionuclide S-35 ($T_{\frac{1}{2}} = 87$ d; $E_\beta^{\max} = 0.169$ MeV) made possible the application of the method of the radioactive surface to the diffusion of sulfur. When the radioactive atoms diffuse into the inner parts of the sample, the measured intensity of the radiation decreases with respect to the increasing absorption of the β-particles. The diffusion coefficient can be determined by such measurements. This method is especially suited to investigations on single crystals where mechanical grinding is difficult. The applied method requires a knowledge of the absorption coefficient of selenium for the β-radiation of S-35. The mass-absorption coefficient for different β-energies was determined by measurement of absorption curves under our special experimental conditions.

The diffusion of sulfur was measured by two modified methods of the radioactive surface, firstly by isothermal step-by-step experiments, and secondly by a continuous procedure according to Nölting.[6]

A thin layer (1 micron) of sulfur labeled with S-35 was deposited on the polished face of the specimen from a carbon disulfide solution. At the beginning the counting rate was about 10^4 cpm. For both methods the decrease of the integral activity was recorded continuously by means of the following experimental set-up (Fig. 2).

The sample was placed in a special socket of a diffusion chamber. The latter was brought to the desired temperature by an ultrathermostat filled with silicon oil. An inert atmosphere (N_2) was maintained during the whole test preventing evaporation and oxidation of the specimen surface. A thermocouple directly under the sample measured the temperature. The β-particles were collimated

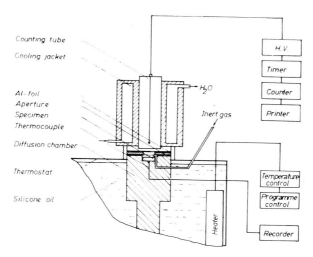

Fig. 2. Experimental set-up.

by an aperture and counted with a cooled Geiger–Müller tube that was separated from the chamber by a thin aluminum foil.

For the case of diffusion from an infinitesimally thin layer into a semi-infinite body the diffusion isotherm for the method of the radioactive surface is given by:

$$F(t) = \frac{I(t)}{I(0)} = \exp \mu^2 Dt \left[1 - \exp(\mu\sqrt{(Dt)})\right]$$

$\frac{I(t)}{I(0)}$ is the ratio between the activities after the diffusion time t and at the beginning. μ is the linear absorption coefficient (1170–1230 cm^{-1}) and D the diffusion coefficient. D was obtained by comparing the measured ratio F with calculated plots of F against $\mu\sqrt{(Dt)}$. For uniform diffusion kinetics an isotherm in a diagram, where $\mu\sqrt{(Dt)}$ is plotted against $\sqrt{t}$, should be a straight line. Figure 3 shows for some measured isotherms that this behavior is approximately realized. The step-by-step method allowed the determination of the diffusion constants for four different temperatures in one run. With the other method according to Nölting all diffusion coefficients in a given temperature range can be obtained, if the heating of the sample is slow and linear. The heating velocity of 3°C/hr was controlled by a programmer. The diffusion coefficients were calculated by a computer.

The results of both methods for the sulfur diffusion can be plotted in the Arrhenius diagram as straight lines (Fig. 4) from which a mean activation energy of $Q = 27$ kcal/mol and a frequency factor of $D_0 = 3.5 \times 10^4$ cm^2/sec as diffusion parameters were calculated. From the analysis of autoradiographs

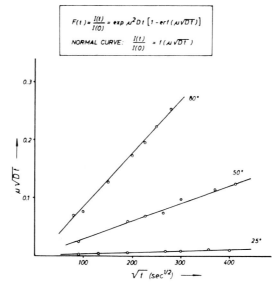

FIG. 3. Kinetic curves.

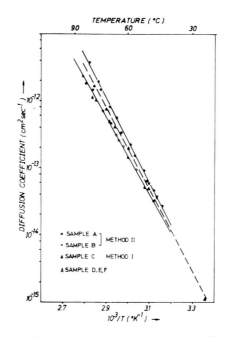

FIG. 4. Diffusion coefficients of S-35 in selenium as a function of temperature.

and the kinetic curves it was assumed that there exists mainly bulk-diffusion in the sample.

In order to obtain more understanding about the diffusion mechanism an investigation has been commenced of the concentration profile through removal of the diffusion region layer by layer.

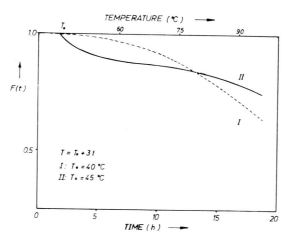

FIG. 5. Recrystallization effect.

The influence of recrystallization effects was tested on the shape of the kinetic curves by the Nölting method. In Fig. 5 the curves are shown of pellets which were annealed (curve I) or not (curve II) after pressing. Curve I shows an increasing decline, whereas curve II is characterized by an inflection point. X-ray studies yielded amorphous and crystalline reflections of the non-annealed sample, whereas the annealed one was fully crystallized. The application of high pressures upon the polycrystalline material thus caused a partial transformation into the amorphous form.

Besides the new analysis methods and the diffusion experiments the crystallization behavior of selenium was studied. Single crystals, grown from $Se + 0.8$ ppm Cl, contained the same amount of chlorine. Single crystals, however, grown from $Se + 615$ ppm Cl, contained only 28 ppm Cl. The crystallization is naturally cleaning the material. The electrical properties of these two single crystals were the same. We must conclude that chlorine in single crystals is not effective.

The influence of impurities on the crystal growth was also considered. Glassy amorphous layers doped with 500 ppm As, Sb and O (addition of SeO_2) were annealed. The radial crystal growth velocity of the obtained spherulites was measured as a function of temperature (Fig. 6). Q is the activation energy of self-diffusion in the amorphous form, ΔG is the difference

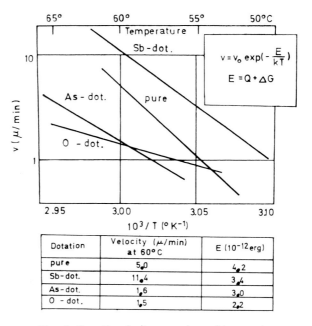

Dotation	Velocity (μ/min) at 60°C	E (10^{-12}erg)
pure	5.0	4.2
Sb-dot.	11.4	3.4
As-dot.	1.6	3.0
O-dot.	1.5	2.2

FIG. 6. Growth velocity vs. reciprocal temperature.

of the free energies of the crystalline and the amorphous phases. The table in Fig. 6 shows average values for the growth velocity at 60°C and average activation energies E for the different growth velocities. The curves are plotted

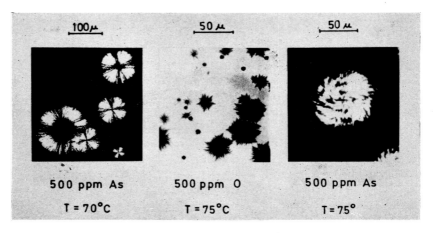

FIG. 7. Crystal shapes of selenium.

according to these values. Oxygen impurities diminish E. If oxygen-doped samples are annealed at 75°C the formation of spherulites can be observed very well (Fig. 7). By annealing the oxygen-free samples at 75°C, a degeneration of the spherulites like in polymers was obtained. With the sample between two polarizers, spirals were seen that can be explained by a wave-formed arrangement of the c-axes. In analogy to the degenerated spherulites of polymers it was assumed that these waved lines represent the projections of spirals.

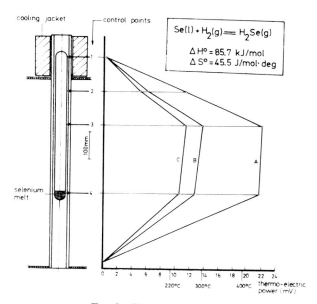

FIG. 8a. Transport reaction.

Among the experiments on the preparation of good selenium single crystals, a chemical transport reaction was investigated. From the pure vapor phase selenium crystals can be grown only with extremely low growth velocities.

An attempt was made to overcome this problem by using a selenium compound, e.g. H$_2$Se. Hydrogen reacts with selenium according to the following equation (Fig. 8a):

$$\mathrm{Se(liquid) + H_2(gas) \rightleftharpoons H_2Se(gas)}$$

The rate of transport depends on the reaction enthalpy ΔH and the reaction entropy ΔS (for H$_2$Se, $\Delta H^\circ_{298} = 9.17$ kcal/mole, and $\Delta S^\circ_{298} = 13.56$ entropy units).[7] The difference of the partial pressures of H$_2$Se at the two different temperatures is a function of the temperatures. Between 473 and 673°K the difference of the partial pressures of H$_2$Se is 10^{-4} mm Hg. This is rather small compared with the partial pressure of selenium that is 1 mm Hg at 600°K. So a

H_2-pressure (mm Hg)	Temperature of the melt (°C)	Growth form	Growth velocity (mm/d)
100–400	430	Whiskers length 2 cm	1.0
30–50	290	Prisms length 1 cm	(0001) 1.0 (10$\bar{1}$0) 0.1

FIG. 8b. Transport reaction.

competition was expected between the transport by H_2Se molecules and by selenium molecules.

Figure 8a shows the experimental set-up. A pumped-off and sealed quartz vessel was used. It was filled with hydrogen of a definite pressure and then placed into a vertical furnace. In the furnace various temperature gradients were maintained during the different experiments to find out the optimal values of the temperatures and the pressure. The growth form and the growth velocity of the crystals turned out to depend on the hydrogen pressure (Fig. 8b).

In order to test whether the transport of selenium from the melt to the surface of the growing crystal took place by the pure selenium or by H_2Se, nitrogen was used as filling gas instead of hydrogen.

Under these conditions a complete change of the growth form was observed. The crystals now had the form of thin platelets of the size 0.5×0.5 cm².

Apart from the transport reaction there is another possibility of reducing the percentage of Se_6 rings in the vapor. Investigations on the selenium molecules of a molecular beam were made by means of a field ionization mass spectrometer. They revealed a surprisingly low number of Se_6 molecules, whereas an extraordinarily high percentage of Se_2 and Se_5 associations was found (vaporization temperature about 170 °C).

An attempt is being made to utilize this fact and to grow single crystals by an arrangement where the vaporizing molecules get to the substratum without having interacted with one another.

To study further the crystallization behavior of selenium, calorimetric measurements are being performed by means of a differential enthalpy calorimeter. It allows the direct determination of transition energies. We measure the difference in electrical power required in maintaining sample and reference at the desired temperature contrary to the well-known differential thermal measurements where the temperature difference is measured.

Vitreous and red amorphous selenium have also been investigated. Vitreous selenium was obtained by the rapid quenching of molten selenium, whereas the red amorphous form is produced by passing selenium vapor first through a temperature region of about 1000 °C and then into liquid nitrogen. The thermograms of freshly prepared amorphous selenium, vitreous or red amorphous, only reveal the peaks due to the crystallization and melting processes. After an annealing time of a few days at a temperature of 24°C the appearance was observed of another peak that grew in the course of time describing an endothermic transition. This peak is due to the slow formation of a state which is decomposed at the glass transition point T_g, experimentally in the temperature range between 35° and 45°C. Our investigations on the red amorphous phase yielded another endothermic peak at a temperature of about

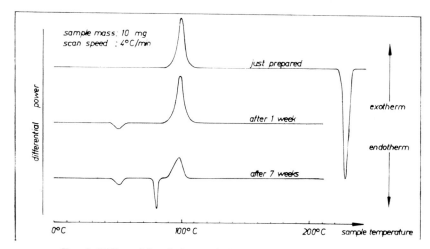

FIG. 9. Differential enthalpy analysis of red amorphous selenium.

70°C. Figure 9 shows a thermogram after a storage time of 7 weeks at a temperature of 24°C. First attempts to interpret this phenomenon as a transition from a certain order into a state of lower order or a break-up of rings respectively failed. The effect could be explained by the dehydration of H_2SeO_3 which is formed according to the following reaction:

$$Se + H_2O + O_2 \rightleftharpoons H_2SeO_3$$

The water and oxygen originate from the surrounding atmosphere. It is known that H_2SeO_3 decomposes into H_2O and SeO_2 at approximately 70°C. Furthermore, the infrared transmission was measured following melting and quenching the red amorphous selenium. An absorption maximum near 11 microns was found which is believed to be caused by SeO_2. By comparison with known mixtures of $Se + H_2SeO_3$, the calculation revealed a content of 2 per cent H_2SeO_3.

The determination of the surface tension of liquid selenium is a further method of studying the influence of impurities. As an example two drops are

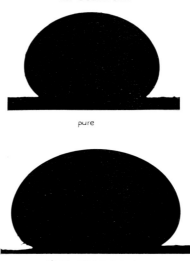

pure

about 400 ppm oxygen

FIG. 10. Drops of liquid selenium.

shown in Fig. 10, one of pure the other of oxygen-doped selenium. It is possible to calculate the surface tension by analyzing the shape of the drop. Figure 11 gives an extract of the formalism necessary for the calculation. The measurements revealed that impurities diminish the surface tension. A particular behavior is shown by copper. It is impossible to have a selenium drop formed on a copper plate since it becomes flat at once.

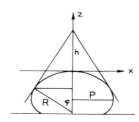

LAPLACE equation of a liquid surface:

$$\sigma\left(\frac{1}{R_1} + \frac{1}{R_2}\right) = \rho g z + C$$

transformed:

$$\frac{1}{r'} + \frac{\sin\varphi}{x'} = 2 + \beta z'$$

surface tension:

$$\sigma = \frac{\rho g b^2}{\beta} = \frac{\rho g P^2}{V^2 p'^2 \beta}$$

FIG. 11. Determination of the surface tension.

Theoretical calculations of the band structure of selenium and tellurium were carried out recently by Madelung and co-workers. These theoretical structures could be examined thus far only in the region of the fundamental transition. The usual optical methods have now been expanded by the electro-reflectance method suggested by Seraphin. The great merit of this new technique is that it is capable of detecting very small changes in the optical properties not only in the region of the fundamental edge, but also at much higher levels in the conduction band.

Electroreflectance is based on the modulation of the optical reflectivity by an applied electric field. This modulation is generally produced in the usual field effect arrangement, the transparent field effect electrode being separated

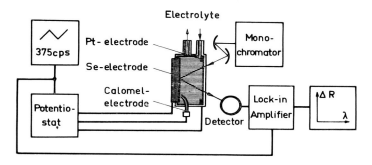

FIG. 12. Block diagram of the experimental set-up.

from the semiconductor surface by a thin mylar spacer. To have a sufficient field modulation in the semiconductor surface, one must apply very high voltages, and it is very difficult to estimate the field strength exactly by surface conductivity measurements.

Therefore, the electroreflectance effect was investigated with another arrangement, viz, the semiconductor–electrolyte interface. The field effect capacitor is built up herein by the very thin Helmholtz layer, and there is no difficulty in realizing a high field strength in the semiconductor surface by a relatively low polarization voltage. As shown by earlier investigations on germanium electrodes, the applied voltage is identical with the potential drop in the space charge layer of the semiconductor under certain conditions (if a special polarization technique is used). The electric field can be determined very exactly by capacity measurements, because the Helmholtz capacitance is large compared to the space charge capacity.

The experimental arrangement is given in Fig. 12. A Cl-doped selenium plate (about 1 mm thick) was used. The electrolyte was pumped with constant flow through the reaction cell. The selenium electrode was potentiostatically

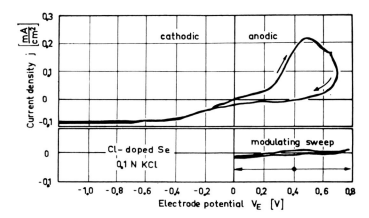

Fig. 13. Current density vs. electrode potential for a Cl-doped Se-electrode in 0.1 N KCl.

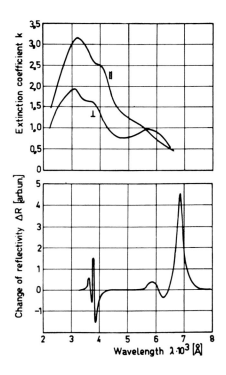

Fig. 14. Optical properties of selenium.

polarized by means of a triangular modulating voltage (375 cps) using a platinum electrode and a calomel reference electrode. Monochromatic light from a Leiss monochromator and a xenon high-pressure lamp is reflected at near normal incidence from the selenium surface and detected by a photomultiplier. The lock-in amplifier beats the multiplier signal against the modulating field voltage feeding the phase-corrected output into a recorder.

The electrochemical behaviour of the selenium–electrolyte interface is described by the current-voltage curve given in Fig. 13. The electrolyte used was 0.1 N KCl. In the cathodic region the characteristic saturation current of the selenium electrode is observed which is due to the limited diffusion of the minority carriers to the surface. For the reflectance measurements the voltage was modulated in the anodic region shown in the lower part of Fig. 13. The polarization amplitude was 800 mV peak to peak. There is no significant current in the working point. Therefore, it was assumed that the voltage modulation is essentially given by the potential variation in the space charge layer of the semiconductor.

There are two characteristic groups of peaks in Fig. 14 showing the differential reflectivity as a function of wavelength. Three peaks are near the band edge at 680, 630 and 600 mμ or at 1.8, 1.95 and 2.05 eV, respectively. Four peaks appear between 410 and 370 mμ or between 3 and 3.3 eV respectively. Both groups can be related to transitions calculated by Olechna and Knox.[8] They agree with the measurements of Chen[9] performed at the polycrystalline Se-1 N Na_2SO_4 interface. Chen's electroreflectance spectrum shows additional peaks between 1.8 and 3.4 eV which have not been found in this laboratory.

It is not possible yet to clear up the origin of all peaks, but it can be seen that the structure is in accordance with our earlier measurements of the extinction coefficient shown in the upper part of Fig. 14. Comparing these two measurements one can see that the resolving power of the new technique is improved essentially. It is hoped to get more information about the band structure of selenium in the near future by applying this new method.

REFERENCES

1. GOBRECHT, H., TAUSEND, A., BRÄTTER, P. and WILLERS, G., *Solid State Comm.* **4**, 307 (1966).
2. GOBRECHT, H., TAUSEND, A., BRÄTTER, P. and WILLERS, G., *Solid State Comm.* **4**, 311 (1966).
3. SAMPATH, P. I., *J. Chem. Phys.* **45**, 3519 (1966).
4. CHEN, I. and DAS, T. P., *J. Chem. Phys.* **45**, 3526 (1966).
5. CHEN, I., *J. Chem. Phys.* **45**, 3536 (1966).
6. NÖLTING, J., *Z. Phys. Chem.* **32**, 154 (1962).
7. RAWLING, J. R. and TOGURI, J. M., *Can. J. Chem.* **44**, 451 (1966).
8. OLECHNA, D. J. and KNOX, R. S., *Phys. Rev.* **140**, A986 (1965).
9. CHEN, J. H., *Phys. Letters* **23**, 516 (1966).

DISCUSSION

CRYSTAL: In the observation of the effect of impurities on the spherulitic growth in selenium, first how thick were your films and second, were they observed with transmitted or incident polarized light?

GOBRECHT: Our films were 2 microns thick and they were observed with transmitted polarized light.

LANYON: Did you find that the previous thermal history of your samples affected the temperature of recrystallization in your DTA work?

GOBRECHT: We can say so far that the homogeneous phase of our specially prepared red amorphous selenium crystallizes at nearly 85°C, whereas the crystallization temperature of vitreous selenium is higher and it is not so reproducible. We are about to investigate the influence of previous thermal history more precisely.

WALLDEN: 1. Regarding surface tension measurements and the influence of oxygen, may 500 ppm O_2 be dissolved as SeO_2? More likely O_2 included in chains influences chain length and viscosity. May viscosity measurements be used as a tool to study chain length and structure in molten selenium?

2. The scientists are using high-purity selenium of different origin with unknown purity and physical structure (unknown thermal history). Suggestion is made that all scientists should have purest possible, uniform selenium available from a "bank".

GOBRECHT: As to our infrared investigations, it is likely that oxygen is dissolved as SeO_2 because we always measured the absorption lines of SeO_2. It cannot be excluded, however, that oxygen is built into the chains, but it is not sure that it will change the average chain length. We have not performed viscosity measurements from which a conclusion might be drawn.

SINGLE-CRYSTAL GROWTH OF HEXAGONAL SELENIUM FROM IMPURITY DOPED MELTS

R. C. KEEZER

Research and Engineering Center, Xerox Corporation, Webster, N.Y.

INTRODUCTION

Monoclinic selenium is an unstable eight-membered ring structure with D_{4d} symmetry. The stable crystalline form of selenium is trigonal[1] (historically called hexagonal) with D_3^4 or D_3^6 symmetry. This crystal structure consists of parallel spiral chains of selenium atoms terminating at the corners and center of a regular hexagon. The bonding within the chains is much stronger than the bonding between the chains.

Although the element has a number of important uses, large selenium single crystals have not been grown until just recently.[2-4]

This paper describes the use of specific impurities which facilitate the growth of selenium from the melt. Normally the presence of impurities is undesirable[5] because growth rates can be limited by the diffusion of impurities from the solid–liquid interface. Constitutional supercooling effects are encountered unless the imposed velocity is reduced below that of the pure material.

The growth rate for pure selenium is very low[6] and diffusion limited growth rates become attractive if the impurity is chosen properly.

Monovalent impurities such as chlorine, bromine, thallium, or potassium appear to simplify the structure of the melt and can be used to obtain selenium single crystals by the Czochralski, traveling solvent, vapor–liquid–solid and gradient freeze techniques.

THE MOLECULAR STRUCTURE OF LIQUID SELENIUM

According to Andrade's[7] theory for monoatomic liquids the viscosity of liquid selenium at the melting point should be about 10^{-2} poise. Since the viscosity of the liquid[8] is several thousand times larger than this value, as shown in Fig. 1, it is concluded that a complex molecular structure is present in the liquid.

The most generally accepted model for the structure of liquid selenium[9] assumes that there is a dynamic equilibrium between eight-membered rings

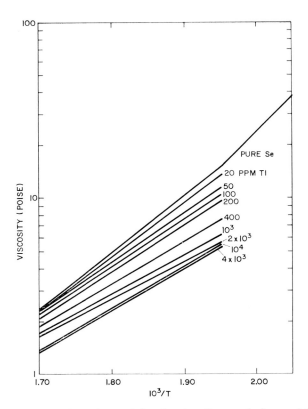

FIG. 1. Temperature dependence of the viscosity of pure selenium and thallium-doped selenium.

(Se_8) and long polymeric chains (Se_n). These species represent the molecular components of the two crystalline forms of selenium, i.e. monoclinic and trigonal. Although this model is probably over-simplified, it appears to explain qualitatively the known behavior of the liquid. The unusually high viscosity is explained by the long chain species[10] and the solubility of the supercooled liquid in CS_2 is attributed to the ring species.[11] The infrared and Raman spectra[12] of the supercooled liquid have been explained by the presence of Se_8 molecules and polymeric Se_n chains. X-ray data on the liquid and supercooled liquid are controversial; however, the existence of both ring and chain species has been reported.[13, 14] Figures 2 and 3 show the available estimates[8] of the weight fraction rings and average chain length in liquid selenium. These estimates indicate that, at the melting point, about 50 per cent of the atoms are in the ring species and the average chain length is between 10^4 and 10^6 atoms/chain.

THE INFLUENCE OF IMPURITIES ON THE MOLECULAR STRUCTURE OF LIQUID SELENIUM

The viscosity of liquid selenium is drastically lowered by the addition of small amounts of monovalent impurities such as chlorine, bromine, potassium, iodine or thallium. Figures 1 and 4 show the influence of chlorine[15] and thallium[8] on the viscosity of liquid selenium. Since the high viscosity is attributed to the long selenium chains, it appears that the average chain

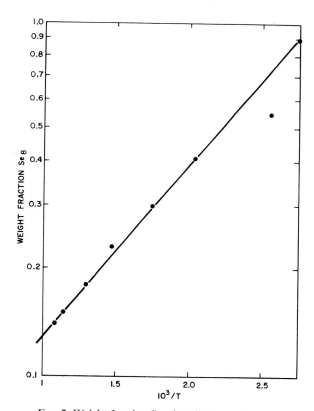

FIG. 2. Weight fraction Se_8 rings in pure selenium.

length is reduced by these impurities. Strongly bonded monovalent elements are expected to terminate free chain-ends and therefore decrease the average chain length when their concentration exceeds the equilibrium chain-end concentration. According to this explanation the viscosity should decrease only after chain-end saturation. Termination of free chain-ends by impurities should reduce the ring concentration by removing chains from the ring-chain

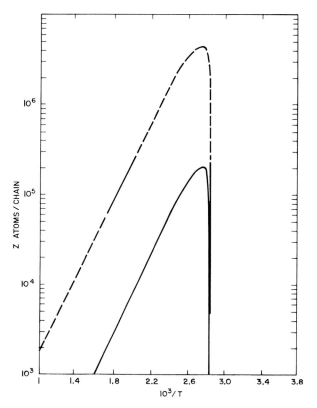

Fig. 3. Chain length in pure selenium. Upper curve Keezer and Bailey.[8] Lower curve Eisenberg and Tobolsky.[9]

equilibrium. Experimental evidence for this last effect is, however, not available. Monovalent impurities are expected to reduce the average chain length whereas other impurities should complicate the liquid structure by chain branching, cross-linking and co-polymerization. This line of reasoning has been used successfully in order to choose selectively several elements which facilitate the growth of hexagonal selenium from the melt.

EFFECT OF IMPURITIES ON CRYSTALLIZATION KINETICS

Pure selenium crystallizes very slowly, as should be expected, due to the complex structure of the liquid.

The high viscosity and related low atomic mobility of liquid selenium preclude rapid crystallization kinetics. Kinetically[16] one expects that if single-crystal growth is possible it should proceed very slowly. Experimentally it is

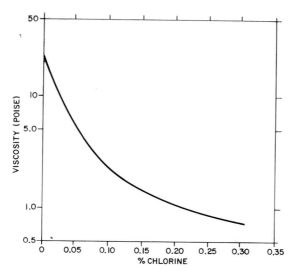

FIG. 4. Viscosity of chlorine-doped selenium (Shirai et al.[15]).

found that liquid selenium can be easily supercooled without crystallization occurring. Henkels and Maczuk[6] have reported that small single crystals can be obtained, after several weeks, by maintaining selenium just below the melting point. Borelius[17] has shown a reciprocal correlation between the crystallization velocity (ω) of selenium and the viscosity (η) of the form $\omega = \text{constant}/\eta$.

Since the viscosity and molecular complexity of liquid selenium are reduced by the addition of monovalent impurities, it should be expected that these impurities can be used to increase the rate of crystallization. Single-crystal growth rates for pure selenium are limited by the mobility of the complex molecular species at the solid–liquid interface. After simplification of these species by the addition of specific impurities, single-crystal growth is primarily limited by the diffusion of the impurity from the solid–liquid interface. The concentration of impurities at the solid–liquid interface may be considerably higher than the average concentration in the liquid and thus the crystallizing liquid may be greatly simplified.

The use of impurities in order to grow single crystals of selenium presupposes partial rejection of the impurity by the crystal and is the situation most commonly encountered. If the impurity is sufficiently soluble in the selenium crystal, growth may not be limited by diffusion but the crystal will contain correspondingly greater quantities of the impurity.

It is also possible that the solubility of the impurity in liquid selenium is too low to accomplish chain-end saturation. This appears to be the situation in the case of silver.

The importance of phase equilibrium data in choosing growth techniques and appropriate growth conditions is best shown by considering the crystallization of selenium from thallium-doped melts.

GROWTH FROM THALLIUM-DOPED MELTS

Thallium usually exhibits monovalent behavior and, as shown in Fig. 1, it reduces the viscosity of liquid selenium. The Se–Tl phase diagram[18, 19] shown in Fig. 5 exhibits a Tl_2Se_3–Se eutectic at 172°C and a liquid-Se monotectic at 201°C. The monotectic contains about 300 ppm thallium. Between

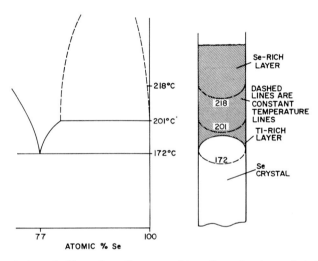

FIG. 5. Selenium–thallium phase diagram and traveling solvent growth technique.

172°C and 218°C selenium should crystallize from a supersaturated selenium–thallium solution. Between 172°C and 201°C the selenium–thallium ratio in the solution is about 8 : 2 and may correspond to a reduction in average chain length to about eight selenium atoms per chain. Selenium single crystals were grown from thallium-doped melts by the vapor–liquid–solid (VLS), traveling solvent (TS), Czochralski and gradient freeze techniques.

The VLS technique[20] makes use of the equilibrium which can be established between a selenium single crystal and a selenium–thallium melt. The melt which usually forms a drop on the tip of the crystal accepts selenium from the vapor and deposits it on the growing crystal. The temperature of the solid–liquid interface must be lower than the vapor–liquid interface in order to insure diffusion of selenium to the growth surface. VLS crystals were grown[21] at about 185°C and at about 210°C. The growth habit was the same at both temperatures even though the liquid drop at the lower temperature should contain about 25 per cent thallium and only a few hundred ppm of thallium at the higher temperature. The growth substrate was either a 1 mm thick polished quartz plate, onto which thallium had been evaporated, or a 2 per cent thallium–selenium alloy. Figure 6 shows the growth morphology of a VLS selenium crystal grown at about 185°C. Tapered platelets growing in a $\langle 10\bar{1}0 \rangle$ direction were obtained in every growth experiment. Hollow, hexagonal prisms, growing parallel to the

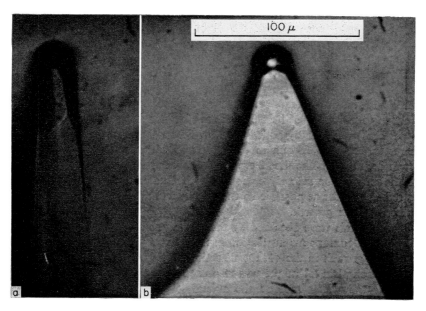

FIG. 6. Vapor–liquid–solid selenium crystal.

c-axis were found when thallium was excluded from the growth tube. Traces of thallium changed the growth habit to the platelet form and caused the growth direction to be perpendicular to the c-axis.

The traveling solvent technique[22] relies on the fact that a liquid alloy will migrate through a solid solvent under the influence of a temperature gradient. The method is analogous to the VLS technique except that the vapor source is replaced by a solid source in contact with the hot side of the liquid alloy. The unusual features of the selenium–thallium phase diagram have made possible the use of a new variation of the TS technique. This method could be called the liquid–liquid–solid technique and establishes exceptionally good growth conditions for selenium.

In the selenium–thallium system as shown in Fig. 5 two liquid layers are formed by establishing a temperature gradient along a selenium–thallium alloy. The phase diagram requires a solid–liquid equilibrium between 172°C and 201°C and a liquid–liquid equilibrium above 201°C. The selenium-rich liquid has a density of about 4 g/cm^3 as compared with a density of 5.8 g/cm^3 for the thallium-rich liquid. The thallium-rich layer is therefore adjacent to the growing crystal. The growth process is also shown schematically in Fig. 5 and has been verified by etching studies on ingots which were quenched during growth. An especially important feature is that the solid–liquid interface is convex toward the liquid if the radial temperature gradient increases toward the center of the ingot. This radial temperature gradient occurs naturally in most growth furnaces and permits a stable concentration gradient only for a convex interface. A convex solid–liquid interface generally minimizes interference by nucleation at the wall of the growth tube. The initial thallium concentration determined the length of the thallium-rich zone. A 15-mm ingot diameter required about 0.2 atomic per cent thallium in order to establish a 1-mm thallium-rich zone. Excess thallium (about 0.5 per cent) was usually added to allow for losses during initial nucleation. Quartz tubes up to 25 mm in diameter and 25 cm in length were used as growth containers. The first 8 cm of the tubes were tapered to about 1 mm and the entire tube severely etched in hydrofluoric acid. The wall thickness was reduced as far as possible so that the quartz tube broke when the crystal was cooled after growth.

Single crystals were obtained by this technique at growth rates as high as 2.5 mm/hr. Spontaneous nucleation in the narrow portion of the tube always resulted in growth in the $\langle 11\bar{2}0 \rangle$ direction.

The Czochralski technique was also used to obtain selenium crystals from the melt. Using this method the crystal is pulled from the selenium-rich layer at about 201°C. Two liquid layers are allowed to form, under argon, in a heavily doped melt (1–5 per cent thallium) and equilibrium is established between a seed and the selenium-rich layer. Growth is obtained by rotating and slowly raising the seed. Centimeter-sized single crystals have been obtained at growth rates as high as 2.5 mm/hr. Growth always took place in the $\langle 11\bar{2}0 \rangle$ direction and attempts to seed in the $\langle 0001 \rangle$ direction always resulted in polycrystalline ingots. Attempts to pull crystals from the selenium rich layer by using a 25 per cent thallium melt were not successful. Rapid dendritic growth was observed and the ingots contained cell growth proceeding in the $\langle 0001 \rangle$ direction.

Large gradient freeze crystals were grown by passing a 1 per cent thallium-doped melt through a horizontal tube furnace. A flat-bottomed quartz boat, tapered on one end, was used to contain the melt. Growth again took place in the $\langle 11\bar{2}0 \rangle$ direction with the $\langle 0001 \rangle$ direction approximately normal to the bottom of the boat.

GROWTH FROM POTASSIUM-DOPED MELTS

The selenium–potassium phase diagram[18] is similar to the selenium–thallium phase diagram. However, the density of the potassium-rich liquid is less than that of the selenium-rich layer. Therefore, the potassium-rich layer floats on the selenium-rich layer and it is not possible to use the liquid–liquid–solid growth technique.

Large selenium crystals have been grown by the vertical Bridgman method from potassium-doped melts. Thin-walled quartz tubes, tapered on one end, were used to contain the melt. The initial potassium concentration was only 0.1 atomic per cent for growth rates of about 1 mm/hr. The crystals grew in the $\langle 11\bar{2}0 \rangle$ direction in a temperature gradient of about 10°C/cm.

One attempt was made to grow selenium from a potassium-doped melt (1.0 atomic per cent) by the Czochralski technique. An ingot containing two large crystallites was obtained and it is expected that this method would also be suitable for selenium single growth.

Visual observations of the fluidity of potassium-doped selenium melts indicate that the viscosity of liquid selenium is drastically reduced by the addition of potassium. Selenium and potassium combine violently and care should be exercised during the initial reaction of the elements.

Attempts to use silver and sodium as dopants to grow selenium crystals by the Bridgman method were totally unsuccessful. The silver–selenium phase diagram[18] does not show a region in which a silver–selenium liquid is in equilibrium with solid selenium so one should not expect selenium crystal growth in this case. The sodium–selenium phase diagram[18] does show such a solid–liquid equilibrium and crystal growth should be possible by the conventional traveling solvent method. Crystal growth experiments using the other Group I elements were not attempted.

CZOCHRALSKI GROWTH FROM HALOGEN-DOPED MELTS

Selenium single crystals were grown from Cl_2- or Br_2-doped melts by the Czochralski technique.[18] The pulling apparatus utilized resistance furnaces with a temperature control of about ±0.2°C. Pulling rates were varied from 0.023 to 0.25 mm/hr. Temperature gradients were varied to achieve optimum growth rates. Rotational rates of 60–100 rpm were necessary in order to prevent build-up of the Cl_2 concentration at the solid–liquid interface. Czochralski crystals of centimeter dimensions were obtained. Single-crystal growth was obtained only in the $\langle 11\bar{2}0 \rangle$ direction. Micro-crystalline ingots were obtained if chlorine was excluded from the growth apparatus.

In order to utilize chlorine effectively, it was necessary to use a cold reservoir to control the Cl_2 concentration in the melt via the equilibrium reactions

$$Se_2Cl_{2(l)} \rightleftharpoons SeCl_{2(g)} + Se_{(s)}$$
$$SeCl_{2(g)} \rightleftharpoons Se_{(l)} + Cl_{2(solution)}$$

Argon gas was used as a carrier to introduce Cl_2 gas into the system. Liquid Se_2Cl_2 and solid selenium, at the reservoir temperature, established an equilibrium $SeCl_2$ pressure above the selenium melt. The $SeCl_2$ pressure determined the Cl_2 concentration in the growth crucible.

The technique used above was also utilized to grow selenium from Br_2-doped melts but I_2 was not used since the corresponding selenium–iodine compounds are not known.

Gradient freeze experiments were conducted with Cl_2, Br_2 or I_2 using the cold reservoir technique. Crystal growth occurred in a thin layer at the surface of the melt when Cl_2 and Br_2 were used but large crystals were not obtained from I_2-doped melts. The electronegativity difference of selenium–iodine is zero in contrast to values for selenium–chlorine of 0.6 and selenium–bromine of 0.4 and I_2 should not attack the selenium–selenium bond as strongly as Cl_2 and Br_2. Single crystal growth took place with the $\langle 11\bar{2}0 \rangle$ parallel to the temperature gradient.

CRYSTAL QUALITY AND RESIDUAL IMPURITIES

Spectrographic and chemical analyses indicate the selenium crystals grown from chlorine- or thallium-doped melts contain from 200 ppm to 400 ppm residual impurities.

Dislocation etch pit studies using 90 per cent HNO_3 revealed 10^4 pits/cm^2.

REFERENCES

1. UNGER, P. and CHERIN, P., *The Physics of Selenium and Tellurium*, Pergamon, New York, 1969, p. 223.
2. HARRISON, D. E., *J. Appl. Phys.*, **36**, 1680 (1965).
3. KEEZER, R., WOOD, C. and MOODY, J. W., *J. Phys. Chem. Solids Suppl.* **1**, 119 (1967).
4. GRIFFITHS, C. H. and SANG, H., *The Physics of Selenium and Tellurium*, Pergamon, New York, 1969, p. 135.
5. TILLER, W. A., *The Art and Science of Growing Crystals*, Chapter 15, J. J. Gilman, Editor, Wiley, New York, 1963.
6. HENKELS, H. W. and MACZUK, J., *J. Appl. Phys.* **25**, 1 (1954).
7. ANDRADE, E. N. DA C., *Phil. Mag.* **17**, 497 and 698 (1934).
8. KEEZER, R. C. and BAILEY, M. W., *Mat. Res. Bull.* **2**, 185 (1967).
9. EISENBERG, A. and TOBOLSKY, A., *J. Polymer Sci.* **46**, 19 (1960).
10. GEE, G., *Trans. Faraday Soc.* **48**, 515 (1952).
11. BRIEGLEB, G., *Z. Phys. Chem.* **A144**, 321 (1929).
12. LUCOVSKY, G. et al., *Solid State Comm.* **5**, 113 (1967).
13. RICHTER, H. and HERRE, F., *Z. Naturforsch.* **13a**, 874 (1958).

14. KREBS, H. and SCHULTZE-GEBHARDT, F., *Acta. Cryst.* **8**, 412 (1955).
15. SHIRAI, T., HAMODA, S. and KOBAYOSKI, K., *J. Chem. Soc. Japan* **84**, 968 (1963).
16. FINDLAY, A., *The Phase Rule*, Dover, New York, 1951.
17. BORELIUS, G., *Arkiv für Fysik* **1**, 305 (1949).
18. HANSEN, M., *Constitution of Binary Alloys*, McGraw-Hill, New York, 1958.
19. KANDA, F. A., Report No. 4, Syracuse University, Syracuse, New York, 1965.
20. WAGNER, R. S. and ELLIS, W. C., *Appl. Phys. Letters* **4**, 89 (1964).
21. KEEZER, R. C. and WOOD, C., *Appl. Phys. Letters* **8**, 139 (1966).
22. TILLER, W. A., *J. Appl. Phys.* **36**, 261 (1965).
23. KEEZER, R. C. and MOODY, J. W., *Appl. Phys. Letters* **8**, 233 (1966).

DISCUSSION

LANYON: Viscosity does not seem to be the only factor affecting growth rate.

KEEZER: Viscosity is used here only as a measure of the molecular complexity of the liquid. The viscosity should decrease as the molecular complexity decreases. Crystallization kinetics are improved, for several reasons, if the liquid is simplified.

GRIFFITHS: (Comment in reply to point raised by Harrison concerning the changes in growth kinetics produced by impurity doping and pressure.) I would like to point out that in the case of both the Harrison and the Keezer crystals the growth mechanism is different from that found in pure selenium crystallized at atmospheric pressure. Extended chain crystallization is taking place rather than the folded chain crystallization which occurs during spherulitic growth. This change appears to be due, in the case of the Harrison crystals, to the shorter chain length and increased mobility at the higher melting point produced by the applied pressure. This allows fractionation of the chains. The crystals show boundaries perpendicular to the c-axis with spacings of the order of magnitude calculated by Keezer and Bailey for the length of selenium chains. Similar structures have also been observed when polyethylene is crystallized under a pressure of 5 kilobars.

It seems, however, that in the case of crystal growth from impurity doped melts, the dopant can be rejected in such a way that the chains can join end to end. Grain boundaries perpendicular to the c-axis were, therefore, not found in the Keezer crystals.

ETCH PIT STUDIES ON SINGLE CRYSTALS OF HEXAGONAL SELENIUM GROWN FROM THE MELT AT HIGH PRESSURES

J. D. Harrison† and D. E. Harrison

Westinghouse Research Laboratories, Pittsburgh, Pennsylvania 15235

INTRODUCTION

Recently Harrison and Sagar[1] reported on the use of bromine–methanol solutions for producing well-defined etch pits in the ($10\bar{1}0$), ($10\bar{1}2$) and (0001) planes of hexagonal selenium. This paper describes some new etch pit studies on these planes and discusses the relationship between etch pits and defects in the crystals. The etch pits were indexed by the use of the standard projections given in Appendix A which were computed by Hudson and Armstrong.

EXPERIMENTAL PROCEDURE

Etching studies were made on single crystals of hexagonal selenium grown from the melt at high pressures by either random nucleation[2] or seeding.[3] A detailed description of the high-pressure device employed was reported earlier for growth by random nucleation.[2] The seeded growth was done as follows: Beads of "ASARCO" 99.999+ per cent pure selenium were placed in a fused silica vial over a seed crystal. The assembly was inserted into the high-pressure device and air was purged from the system by evacuation and back filling with argon. Using argon as the working fluid, the system was pressurized at from 4000 to 6000 atm. After the system reached steady-state heat flow, the thermal gradient was moved downward until the melting isotherm lay within the selenium seed crystal. The thermal gradient was then moved upward at from 1 to 5 mm/hr, which caused the seed to grow. At the completion of progressive solidification, the heaters were adjusted so that the whole crystal was at a uniform temperature of from 150° to 200°C for several hours. After annealing, the pressure was decreased to 1 atmosphere over a period of from 6 to 60 hr, and the bomb was cooled to room temperature. Upon removal from the high-pressure device, the sample vial was placed in hydrofluoric acid to etch away the fused silica container.

† Present address: Raychem Corp., 300 Constitution Drive, Menlo Park, California 94025.

The ($10\bar{1}0$) and ($10\bar{1}2$) surfaces were revealed by cleavage, whereas the (0001) plane was prepared by cutting the crystal with a S.S. White Airbrasive unit followed by grinding and polishing. Since the $\{10\bar{1}0\}$ planes form the primary cleavage system, $\{10\bar{1}2\}$ cleavages were difficult to obtain. Successful ($10\bar{1}2$) cleavage was assured, however, by first embedding the crystal in an epoxy mounting compound, then cutting a groove through the epoxy encircling the ($10\bar{1}2$) plane, and finally securing one end of the mounted crystal in a vice and striking the free end with a hammer.

The etchant was made from 5 to 10 vol. per cent analytical reagent bromine dissolved in methanol. After etching, the selenium samples were washed with fresh methanol, then dried with a hot-air blower. Etching times of several minutes caused no staining.

EXPERIMENTAL RESULTS

(a) *Growth Morphology*

Single crystals 1 cm in diameter by 10 cm long with the long axis parallel to either the $\langle 0001 \rangle$ or the $\langle 11\bar{2}0 \rangle$ direction were grown from the melt using pre-oriented seed material. Efforts to seed growth normal to the ($10\bar{1}2$) face were unsuccessful. Since the $\langle 0001 \rangle$ seeds were easily prepared while the $\langle 11\bar{2}0 \rangle$ seeds were difficult to make, only one crystal was grown in the $\langle 11\bar{2}0 \rangle$ direction. Of the twenty-one crystals grown in the $\langle 0001 \rangle$ direction, all had a polygranular zone above the seed (Fig. 1). In all but three cases, one grain finally dominated to give a single crystal whose crystallographic orientation agreed exactly with the seed. The seed-to-overgrowth continuity was indicated by the simultaneous reflection of the seed, the melt-grown and the vapor-grown crystal sections. A characteristic of the $\langle 0001 \rangle$ crystals was a terminal surface decorated with a forest of vapor-grown needles. Each needle exhibited six specular ($10\bar{1}0$) faces. Continuity between the seed, the melt-grown and the vapor-grown sections of the crystal was confirmed by Laue back-reflection patterns.

The seeded crystals exhibited smoother, more extensive cleavage on the ($10\bar{1}0$) planes than those which were randomly nucleated. Prismatic cleavage wafers less than 0.62 mm thick transmitted red light detectable by eye in a darkened room when illuminated with an 18 watt Bausch and Lomb Nicholas Illuminator. Regions of mechanical damage appeared as dark shadows. Most samples thinner than 0.25 mm were translucent, but only about half of those tested between 0.25 and 0.62 mm transmitted red light.

(b) *Included Gas*

Some of the argon which dissolved in the liquid selenium was rejected during solidification. As a consequence, gas bubbles as large as 4 mm in diameter formed at the container–crystal–liquid interface leaving hemispheri-

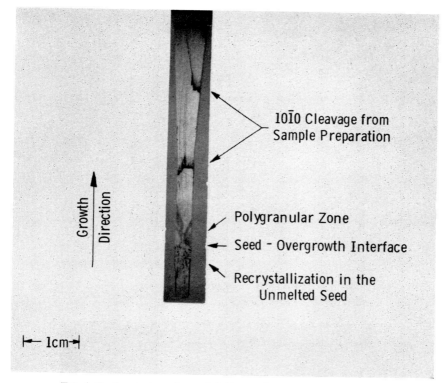

FIG. 1. Sectional view of a seeded 10-mm diameter selenium crystal.

cal marks on the surface of some crystals. In addition, small gas bubbles (from $\frac{1}{20}$ to $\frac{1}{4}$ micron in diameter) formed inside of the 10-mm diameter crystals. Electron micrographs of a replica of a $(10\bar{1}2)$ cleavage surface revealed that the outer 2 mm of the crystal was virtually free of bubbles while the density of bubbles in the interior was of the order $10^5/cm^2$. In addition to isolated spheres, hollow pipes which were elongated in the $\langle 0001 \rangle$ direction were observed in the $(10\bar{1}0)$ cleavage plane. Electron micrographs of replicas of the $(10\bar{1}2)$ cleavage surface revealed the presence of hollow pipes as well as isolated spheres. Neither bubbles nor pipes were detected inside of the 3-mm diameter crystals.

(c) *Etch Figures*

The $(10\bar{1}0)$ prism surfaces of the 10-mm diameter crystals exhibited three types of etch figures: small, diamond-shaped pits (Fig. 2); large elongated grooves (Fig. 3); and etching steps with scalloped edges (Fig. 3). One diagonal of the diamond-shaped pits was aligned in the $\langle 0001 \rangle$ direction and the pits

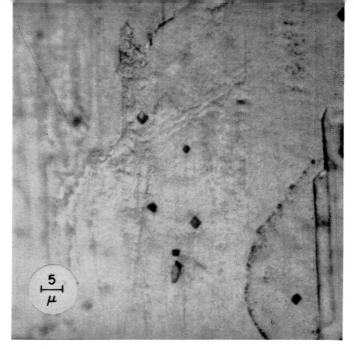

FIG. 2. (10$\bar{1}$0) prism cleavage surface; $\langle 0001 \rangle$ direction vertical; etched in 5 vol. % Br in CH$_3$OH for 8 sec; 10-mm diameter seeded crystal.

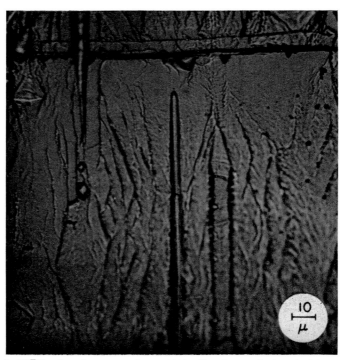

FIG. 3. (10$\bar{1}$0) prism cleavage surface; $\langle 0001 \rangle$ direction vertical; etched in 10 vol. % Br in CH$_3$OH for 4 min; $\simeq 100\ \mu$ removed from surface by etching.

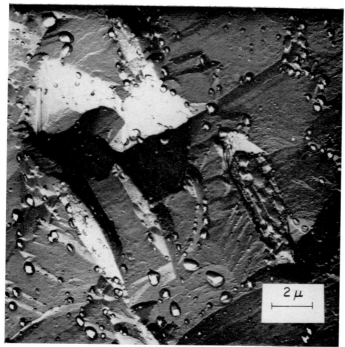

FIG. 4. Electron micrograph of a two-stage carbon replica (shadowed with WO_2 at 45° inclination) from a (10$\bar{1}$0) cleavage surface etched in 5 vol. % Br in CH_3OH for 30 sec.

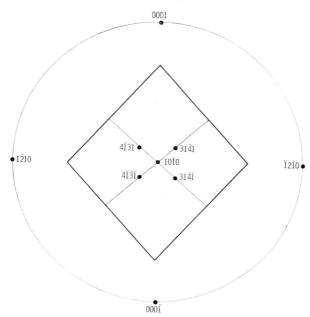

FIG. 5. (10$\bar{1}$0) pole figure of etch pits in hexagonal selenium.

(a)

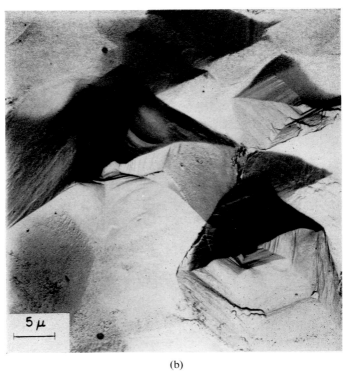

(b)

FIG. 6(a) and (b). Electron micrographs of a two-stage carbon replica (shadowed with WO_3 at 60° inclination) from a (0001) section etched in 5 vol. % Br in CH_3OH for 12 sec.

were shallow (Fig. 4). Since the alternate included angles were 85° and 95°, reference to the standard (10$\bar{1}$0) projection (see Appendix A) indicated that the diamond-shaped pits were bounded by {4$\bar{1}$31} planes as shown by the pole figure in Fig. 5. The elongated grooves which had parallel sides and pointed ends were aligned in the ⟨0001⟩ direction. As the (10$\bar{1}$0) surface etched away, old grooves eroded away and new grooves appeared in different locations (see Fig. 3). By contrast, the (10$\bar{1}$0) cleavage surface of the 3-mm diameter crystals exhibited neither pits nor grooves.

Newly cut, polished and etched (0001) basal plane sections exhibited six-sided pits.[1] Both shallow and deep pits were observed as shown by the electron micrographs in Fig. 6. All of the pits on a particular section had the same geometry, but the pits differed slightly from section to section. The alternate included angles were always 105 and 135 degrees, but the ratio of the length of the short to long sides varied from 1 : 1.4 (see Fig. 7) to 1 : 1 (see Fig. 8). By use of the geometric data and the standard (0001) projection (see Appendix A), the six-sided pits were shown to be formed by the (30$\bar{3}$1) ($\bar{3}$301) (0$\bar{3}$31) and (13$\bar{4}$1) ($\bar{4}$131) (3$\bar{4}$11) planes. This geometry accounts for the observed variation in the ratio of the sides by assuming that the two sets of planes etch at different rates. A pole figure of the 1 : 1.4 and 1 : 1 geometries is shown in Fig. 9. Although the {30$\bar{3}$1} and {13$\bar{4}$1} planes do not converge at a common point, the poles differ by only 2° 20′ from the ⟨0001⟩ direction.

Freshly cleaved and etched (10$\bar{1}$2) pyramid planes displayed six-sided pits which, when graphically projected onto the (0001) surface, produced the etch-pit geometry found on the basal plane. All of the six-sided pits on a particular surface had the same geometry, but differences were observed from specimen to specimen. For example, the six-sided pits in the (10$\bar{1}$2) face of the 10-mm diameter crystal in Fig. 15 had straight sides, whereas the sides of the pits in the (10$\bar{1}$2) surface of a 3-mm diameter crystal were curved (Fig. 16). Electron micrographs established that some of the pits had flat bottoms while others terminated in long pipes.

(d) *Annealing Effects*

An unexpected result was found upon etching a (0001) section of a 3-mm diameter crystal which had been polished 4 years earlier. Instead of six-sided pits, triangular etch pits which were bisected by the ⟨11$\bar{2}$0⟩ direction (Fig. 10) appeared in the basal plane. By means of a two-circle goniometer and the (0001) standard projection, the pits were shown to be formed by the (2$\bar{1}$$\bar{1}$3) ($\bar{1}2\bar{1}$3) ($\bar{1}$$\bar{1}$23) planes. A pole figure of the triangular etch pits in relation to the six-sided pits is shown in Fig. 9. Measurements with a two-circle goniometer also revealed that the etching grooves in the ⟨11$\bar{2}$0⟩ directions were formed by the {10$\bar{1}$2} planes. After repolishing and etching, the same general distribution of pits reappeared but now the pits were the six-sided {30$\bar{3}$1}

FIG. 7. (0001) basal section prepared by cutting, grinding, and polishing; etched in 5 vol. % Br in CH_3OH for 60 sec; 10-mm diameter seeded crystal.

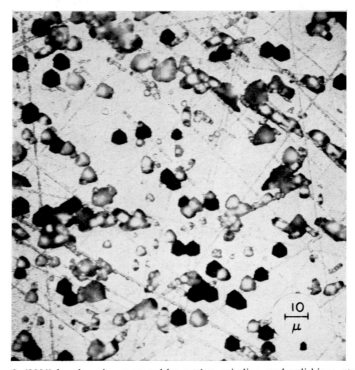

FIG. 8. (0001) basal section prepared by cutting, grinding, and polishing; etched in 5 vol. % Br in CH_3OH for 13 sec; 10-mm diameter seeded crystal.

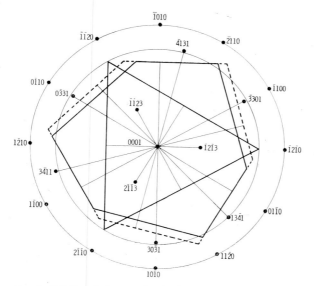

FIG. 9. (0001) pole figure of etch pits in hexagonal selenium.

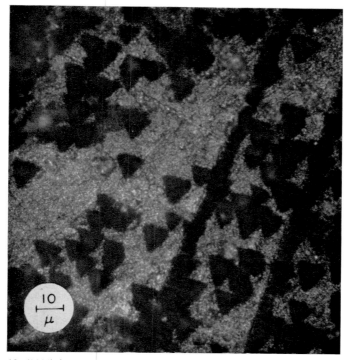

FIG. 10. (0001) basal section prepared by cutting, grinding and polishing; after 4 years surface was etched with 5 vol. % Br in CH₃OH for 80 sec; grooves are in the $\langle 11\bar{2}0 \rangle$ direction; 3-mm diameter crystal grown by random nucleation.

FIG. 11. (0001) basal section prepared by grinding and polishing; after 4 years surface was etched with 5 vol. % Br in CH$_3$OH for 80 sec; 3-mm diameter crystal grown by random nucleation.

FIG. 12. Repolished and etched (0001) basal section shown in Fig. 11; etched with 5 vol. % Br in CH$_3$OH for 8 sec.

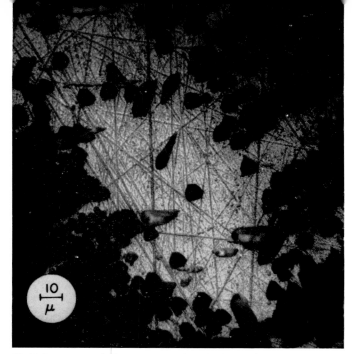

FIG. 13. Enlargement of Fig. 12 to reveal similarity to six-sided etch pits shown in Fig. 7.

FIG. 14. Second repolishing and etching of (0001) basal section shown in Fig. 11; etched with 5 vol. % Br in CH_3OH for 8 sec.

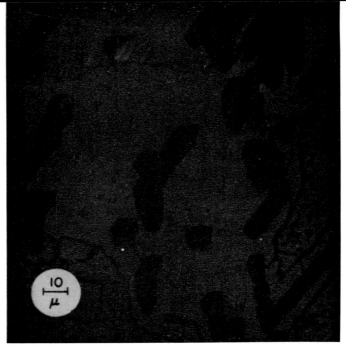

Fig. 15. (10$\bar{1}$2) cleavage face etched in 5 vol. % Br in CH$_3$OH for 5 sec; 10-mm diameter seeded crystal.

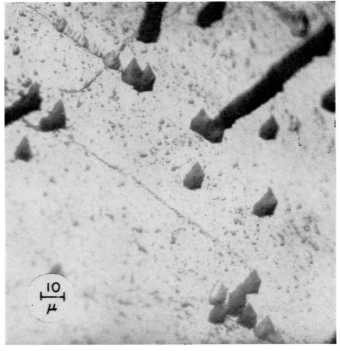

Fig. 16. (10$\bar{1}$2) cleavage face etched in 5 vol. % Br in CH$_3$OH for 5 sec; 3-mm diameter crystal grown by random nucleation.

FIG. 17. (10$\bar{1}$2) cleavage surface etched after one year with 5 vol. % Br in CH$_3$OH for 5 sec; 10 mm diameter seeded crystal.

{13$\bar{4}$1} ones. Figure 11 gives the distribution of triangular pits and Fig. 12 displays the distribution of six-sided pits obtained after repolishing. Figure 13 is an enlargement of Fig. 12 to illustrate that the pits are the same as those produced in the (0001) plane of the 10 mm crystal (Fig. 7). Repolishing and etching a third time again gave the same general distribution of etch pits (Fig. 14). Electron micrographs of replicas revealed that the pits terminated in long pipes similar to that shown in Fig. 6(b).

The effect of aging was investigated by etching a 1-year-old (10$\bar{1}$2) cleavage fragment from the same crystal that gave Fig. 15. Irregular triangular etch pits which were bisected by the $\langle 11\bar{2}0 \rangle$ direction formed (see Fig. 17) instead of the six-sided ones. Graphic projection of the triangular pits from the (10$\bar{1}$2) to the (0001) plane produced the {2$\bar{1}\bar{1}$3} etch pit geometry.

DISCUSSION

The diamond-shaped etch pits and the elongated grooves seen on the (10$\bar{1}$0) cleavage faces of the 10-mm diameter crystals appeared to be the sites of isolated gas bubbles and gas pipes, respectively, particularly since these

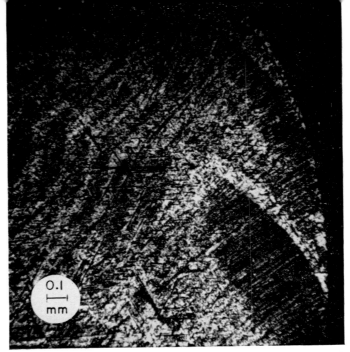

FIG. 18. (0001) basal section prepared by grinding and polishing; etched with 5 vol. % Br in CH_3OH for 8 sec; the rows of etch pits are in the $\langle \bar{1}100 \rangle$ direction; 3-mm diameter crystal grown by random nucleation.

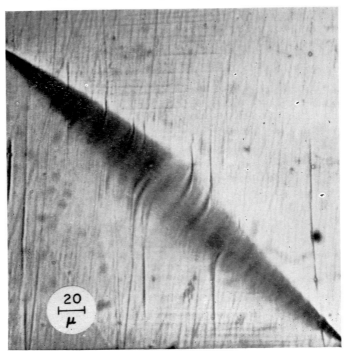

FIG. 19. Polished ($10\bar{1}0$) cleavage face; indentor wedge at 45° to $\langle 0001 \rangle$; ripple marks are parallel to and the cracks are perpendicular to $\langle 0001 \rangle$ direction; 3-mm diameter crystal grown by random nucleation.

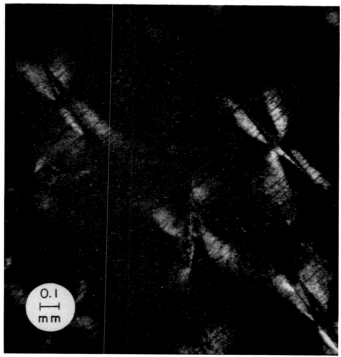

Fig. 20. Polished (10$\bar{1}$0) cleavage face; etched with 5 vol. % Br in CH$_3$OH for 10 sec; dark field illumination reveals preferential etching; 3-mm diameter crystal grown by random nucleation.

features were absent in the 3-mm diameter crystals and scarce in the outer 2 mm of the 10-mm diameter crystals. Elongated bubbles are frequently encountered during the freezing of water, when pipes are formed by bubble caps which move with the ice-water interface.[4]

Undoubtedly, some of the etch pits seen in the (0001) surfaces of the 10-mm diameter crystals were related to gas bubbles, but the majority must have originated from other causes since they were 10 to 100 times more numerous than gas bubbles in the (10$\bar{1}$0) plane. This conclusion is supported by the fact that an etch pit density of $\simeq 10^7/\text{cm}^2$ was found in the (0001) sections of 3-mm diameter crystals in which no gas bubbles were detected. A portion of the etch pits may have resulted from localized surface damage. The pair of mating (10$\bar{1}$2) cleavage surfaces reported earlier[1] appear to show this effect. After progressive etching only about 70 per cent of the six-sided pits could be correlated from one face to the other. The majority of these etch pits, however, appear to be due to grown-in dislocations. Evidence for dislocation motion during growth in randomly nucleated crystals was exhibited by the rows of etch pits in the $\langle \bar{1}100 \rangle$ direction separated by etch-pit-free growth bands which persisted after repeated grinding and polishing (see Fig. 18). Since dislocation motion in a plane requires that the Burgers vector lies in that plane, the indicated slip system is either $\langle \bar{1}100 \rangle$ $\{11\bar{2}0\}$ or $\langle 0001 \rangle$ $\{11\bar{2}0\}$ compared to $\langle 11\bar{2}0 \rangle$ $\{10\bar{1}0\}$ for isostructural tellurium at room temperature.[5]

Localized strain energy which could be annealed out over long periods of time at room temperature (22°C) was revealed by the different etch-pit geometries obtained on aged versus freshly worked surfaces. Triangular etch pits similar to those seen in the present investigation were found by Eckart using a HNO_3-H_2SO_4 etchant at 150°C on the (0001) face of vapor-grown crystals.[6] He also found six-sided etch pits at 20°C in the (0001) plane of crystals grown from chlorine-doped selenium vapor. These results suggest that localized strain caused by either mechanical deformation or impurities lead to the six-sided etch pits, while the triangular pits represent a strain free surface.

No evidence was found for slip bands in the $(10\bar{1}0)$ surfaces as a result of mechanical deformation at room temperature. The orthogonal cracks and ripple marks (see Fig. 19) produced by the Knoop hardness indentor inclined at 45° to the $\langle 0001 \rangle$ direction appeared to involve fracture and plastic deformation respectively. After etching, dark-field illumination showed preferential attack (Fig. 20), but the effect did not reappear after repolishing and etching. If slip bands were present as a result of plastic deformation, they should persist through repolishing and give rise to dislocation etch pits. The absence of dislocation etch pits in the $(10\bar{1}0)$ plane, however, may be due to the insensitivity of the etchant on this surface.

ACKNOWLEDGMENT

The authors are indebted to Mr. J. E. Hudson and Dr. R. W. Armstrong for preparing the standard projections, to Miss Jean Coles for the electron photomicrographs and to Dr. F. F. Lange for his helpful discussions.

REFERENCES

1. HARRISON, J. D. and SAGAR, A., *J. Appl. Phys.* **38**, 3791 (1967).
2. HARRISON, D. E. and TILLER, W. A., *J. Appl. Phys.* **36**, 1680 (1965).
3. HARRISON, J. D., *J. Appl. Phys.* **39**, 3672 (1968).
4. CHALMERS, B., *Scientific Am.* **200**, 114 (1959).
5. STOKES, R. J., JOHNSTON, T. L. and LI, C. H., *Acta Met.* **9**, 415 (1961).
6. ECKART, F., *Recent Advances in Selenium Physics*, Pergamon, London, 1965.

APPENDIX A

STANDARD PROJECTIONS AND CRYSTALLOGRAPHIC ANGLES FOR HEXAGONAL SELENIUM

J. E. HUDSON and R. W. ARMSTRONG†

Westinghouse Research Laboratories, Pittsburgh, Pennsylvania 15235

The following communication presents for metallic hexagonal selenium; first, the (0001) and the (10$\bar{1}$0) standard stereographic projections and, secondly, some tabulated angles between the (0001) or (10$\bar{1}$0) and other plane poles. This information was accumulated during an investigation of the deformation and fracture modes exhibited in selenium single crystals and

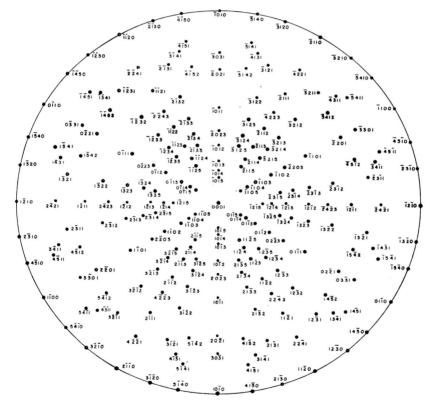

FIG. 1A. Std. (0001) projection for hexagonal selenium (c/a 1.1345).

† Present address: Brown University, Providence, R.I.

TABLE 1. *Calculated Angles Between Planes for Hexagonal Selenium*
(*Hexagonal Indices, c/a = 1.1345*)

$h_1k_1i_1l_1 = 0001$		$h_1k_1i_1l_1 = 10\bar{1}0$	
$h_2k_2i_2l_2$	ϕ	$h_2k_2i_2l_2$	ϕ
$10\bar{1}5$	14° 41′	$41\bar{5}0$	10° 53′
$10\bar{1}4$	18° 08′	$21\bar{3}0$	19° 06′
$10\bar{1}3$	23° 35′	$11\bar{2}0$	30° 00′
$10\bar{1}2$	33° 14′	$12\bar{3}0$	40° 54′
$20\bar{2}3$	41° 08′	$14\bar{5}0$	49° 06′
$10\bar{1}1$	52° 38′	$01\bar{1}0$	60° 00′
$20\bar{2}1$	69° 07′	$\bar{1}5\bar{4}0$	70° 54′
$30\bar{3}1$	75° 43′	$\bar{1}3\bar{2}0$	79° 06′
$10\bar{1}0$	90° 00′	$\bar{1}2\bar{1}0$	90° 00′
$41\bar{5}2$	71° 34′	$22\bar{4}1$	32° 15′
$41\bar{5}1$	80° 33′	$12\bar{3}1$	43° 25′
$41\bar{5}0$	90° 00′	$14\bar{5}2$	51° 36′
		$02\bar{2}1$	62° 09′
$31\bar{4}1$	78° 03′	$\bar{1}2\bar{1}1$	90° 00′
$21\bar{3}5$	34° 44′	$41\bar{5}1$	14° 23′
$21\bar{3}4$	40° 54′	$31\bar{4}1$	18° 15′
$21\bar{3}3$	49° 07′	$21\bar{3}1$	24° 47′
$21\bar{3}2$	60° 01′	$11\bar{2}1$	37° 35′
$21\bar{3}1$	73° 54′	$12\bar{3}2$	49° 06′
$21\bar{3}0$	90° 00′	$01\bar{1}1$	66° 35′
		$\bar{1}3\bar{2}3$	81° 47′
$11\bar{2}5$	24° 25′	$\bar{1}2\bar{1}2$	90° 00′
$11\bar{2}4$	29° 34′		
$11\bar{2}3$	37° 06′	$41\bar{5}2$	21° 19′
$11\bar{2}2$	48° 36′	$21\bar{3}2$	35° 04′
$22\bar{4}3$	56° 32′	$11\bar{2}2$	49° 29′
$11\bar{2}1$	66° 13′	$12\bar{3}4$	60° 20′
$22\bar{4}1$	77° 34′	$01\bar{1}2$	74° 06′
$11\bar{2}0$	90° 00′	$\bar{1}2\bar{1}4$	90° 00′

large-grained polycrystals. Figures 1A and 2A were made using the crystallographic data obtained at 26°C by Swanson, Gilfrick and Ugrinic,[1]

$$a = 4.3574 \text{ kX} \qquad c = 4.9436 \text{ kX}$$

and by calculating the angle between crystallographic planes denoted by hexagonal indices,[2] as

$$\cos \phi = \frac{h_1 h_2 + k_1 k_2 + (1/2)(h_1 k_2 + h_2 k_1) + 3a^2 l_1 l_2 / 4c^2}{\sqrt{[(h_1^2 + k_1^2 + h_1 k_1 + 3a^2 l_1^2/4c^2)(h_2^2 + k_2^2 + h_2 k_2 + 3a^2 l_2^2/4c^2)]}}$$

The values of ϕ, calculated to within 0.5′, are given in Table 1. Although calculated values of the structure factor showed the hexagonal symmetry of the unit cell[3] (e.g. the six pyramidal planes $\{10\bar{1}1\}$ gave two groups of planes

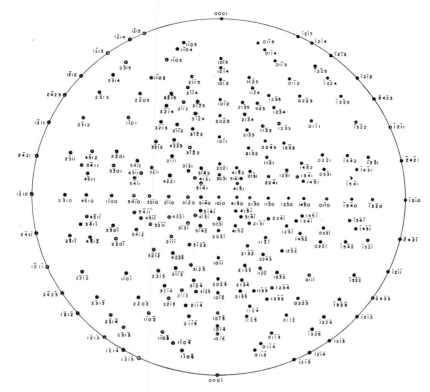

Fig. 2a. Std. (10$\bar{1}$0) projection for hexagonal selenium (c/a 1.1345).

having intensities of scattering differing by a factor of approximately 8) no obvious correlation between calculated and observed intensities occurred in Laue patterns, as might be expected, and the structure factor data have thus been omitted.

ACKNOWLEDGMENT

This work was partially supported by the Air Force Office of Scientific Research under Contract AF 49-(638)-1177.

REFERENCES

1. H. E. Swanson, N. T. Gilfrick and G. M. Ugrinic (1955), U.S. Nat. Bur. Stds. Circ. Nos. 539, **5**, 55.
2. C. S. Barrett, *Structure of Metals*, p. 635, McGraw-Hill, New York (1952).
3. A. J. Bradley *Phil. Mag.* **48**, 477 (1924).

DISCUSSION

Laudise: Did you study perfection with the Schulz technique?

Harrison: No.†

Lanyon: There is apparently a large difference between the plastic deformation of your crystals and those of Keezer. Could this be due to the fact that the chain length is permanently shortened by the addition of impurities whereas in your case on the removal of pressure and reduction of temperature, the chain length will increase causing the brittleness that you observe in contrast to Keezer's plastic flow results?

Harrison: Both the mechanical properties and your results on trapping levels indicate that impurities are associated with defects (presumably dislocations) which modify the structure, possibly by the mechanism you have suggested. Since no evidence was found for dislocation motion at room temperature in "pure" selenium, the ductility exhibited by the thallium-doped crystals implies that thallium modifies the structure to permit dislocation mobility at room temperature. In a similar way your results infer that thallium or potassium is associated with defects since the acceptor concentration for "pure" and doped selenium was of the order $10^{14}/cm^3$ whereas the concentration of thallium or potassium was of the order $10^{18}/cm^3$.

Champness: Can you explain the $(10\bar{1}2)$ cleavage and did cleavage occur in this plane in crystals grown by Dr. Keezer?

Harrison: Primary cleavage along the $\{10\bar{1}0\}$ planes is consistent with the anisotropy in the bond strengths, whereas secondary cleavage on the $(10\bar{1}2)$ planes is puzzling since this requires fracture of the selenium chains. These results do indicate, however, that the $\{10\bar{1}2\}$ family has the lowest surface energy for planes which intersect the $\langle 0001 \rangle$ direction. Electron micrographs revealed that cleavage along the pyramid planes had progressed in a stable mode. As I understand Dr. Keezer's results, he did not observe cleavage on the $(10\bar{1}2)$ planes. In this respect as well as in other properties, the thallium-doped crystals are different from the "pure" pressure-grown crystals.

Siemsen: Did you observe large areas with low angle grain boundaries, as could be seen by simple light microscope techniques with polarized light?

Harrison: The low angle (3–4°) boundaries orthogonal to the $\langle 0001 \rangle$ direction commonly occurring in the $(10\bar{1}0)$ plane of the randomly nucleated crystals were absent in the seeded crystals as revealed by light microscopy.

THE GROWTH OF SELENIUM SINGLE-CRYSTAL FILMS

C. H. GRIFFITHS† and H. SANG‡

Noranda Research Centre, Pointe Claire, Quebec, Canada

INTRODUCTION

The growth of trigonal selenium single crystals from an undoped melt is very difficult due to its highly polymeric nature (chain length approximately 10^5 atoms[1]), and in order to achieve any degree of success it has generally been necessary to reduce the chain length in some way or increase the mobility.[2,3] In the vapor, on the other hand, the polymerization is much lower (Se_2 to Se_8) and isolated crystals can be grown more easily. These crystals, however, normally take the form of small hollow hexagonal needles[4] which are far from being ideal specimens for physical measurements.

Selenium films have been prepared from the vapor by a number of workers but there have been few attempts to produce crystalline films with accurately controlled orientation. Although selenium was shown to grow epitaxially on alkali halides,[5] multiple, rather than single, orientations of the lattice were formed in all cases.

The choice of a substrate material for the preparation of single-crystal selenium films by epitaxial growth is not a simple matter. Attempts have been made to predict the occurrence of epitaxial growth on the basis of misfit between substrate and overgrowth lattices and the bonding across the interface. For a number of reasons these have not been notably successful. In this work the structures of selenium films deposited on ($10\bar{1}0$) and (0001) tellurium and (0001) bismuth telluride are described. Tellurium has the same structure as selenium and similar lattice constants and is therefore an obvious choice. Bismuth telluride cleaves readily in the (0001) plane and the structure is such that the exposed plane is composed only of an hexagonal array of tellurium atoms, as is the tellurium (0001) plane. The bonding perpendicular to the Bi_2Te_3 (0001) plane is, however, Van der Waals rather than the much stronger covalent bonding in the tellurium.

---

† Present address: Research and Engineering Center, Xerox Corporation, Webster, N.Y.
‡ Present address: Department of Metallurgy and Materials Science, University of Toronto, Toronto 5, Ontario, Canada.

EXPERIMENTAL

The selenium films were prepared in a conventional vacuum evaporation chamber shown in Fig. 1. Substrate crystals were mounted on the heater about 10 cm from the vapor source and the temperature of the substrate face was monitored and controlled by a thermocouple sprung onto the surface.

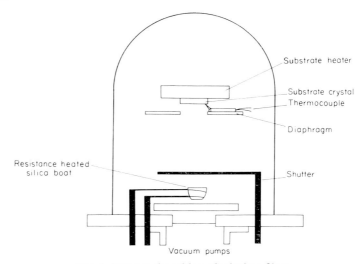

FIG. 1. Vacuum deposition of selenium films.

The selenium was commercial high-purity material from Canadian Copper Refiners Limited and was vacuum distilled before use. It was evaporated from a resistance heated fused silica boat which maintained the selenium at 250–270 °C. A shutter and a diaphragm were mounted between the source and the substrate crystal to control the exposure time and shield the substrate from contamination.

The tellurium crystals were obtained as Czochralski grown ingots and were cut with an abrasive particle jet as shown in Fig. 2. Wafers with large (0001) faces were cut directly in this way, and ($10\bar{1}0$) plane substrates were cleaved from thicker wafers. Selenium was deposited on cleaved ($10\bar{1}0$) faces and on polished ($10\bar{1}0$) and (0001) faces. The substrates were polished in the usual manner using stannic oxide powder as the final polish and were annealed in vacuum for 2 hr at 200 °C immediately prior to the deposition of selenium. This process restored the monocrystallinity of the surface as shown by the Laue back reflection X-ray pattern of a ($10\bar{1}0$) face in Fig. 3.

Films were normally deposited at a rate in the order of 1000 Å per minute to give thicknesses up to about 10 microns. Film thicknesses were determined

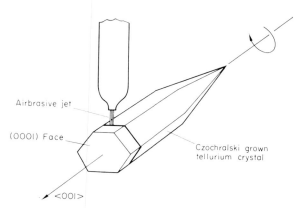

FIG. 2. Tellurium substrate preparation.

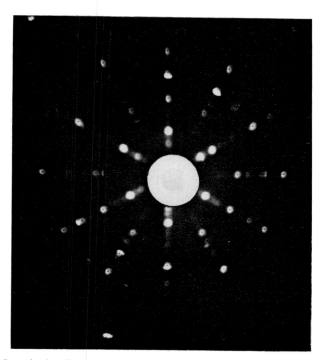

FIG. 3. Laue back reflection pattern from polished and annealed (10$\bar{1}$0) tellurium.

by chemical analysis[5] in the case of films thinner than 1 micron and from infrared interference fringes in the case of films thicker than 1 micron. The influence of substrate temperature on film structure was investigated for films up to 3000 Å in thickness which were thin enough to be examined in the electron microscope. In order to free the films from the substrates for structural examination the tellurium was dissolved away with a 20 per cent aqueous solution of nitric acid. It was found necessary, particularly in the case of the thin (0001) films, to add ferric ions to the solution to inhibit attack of the selenium. Due to the extremely fragile nature of the films it was also found necessary to strengthen them with nitrocellulose or silicone rubber before substrate dissolution. The thinner films were examined in an A.E.I. EM6G electron microscope and the thicker films were subjected to X-ray diffraction.

RESULTS

(a) $(10\bar{1}0)$ *Tellurium Substrate*

Thin selenium films were deposited on cleaved $(10\bar{1}0)$ tellurium at 40–140°C and epitaxial growth in parallel orientation to the substrate was observed from 70° to 135°C. Above 135°C, no selenium deposited on the substrate. Good single-crystal growth occurred between 95° and 135°C and a typical film of this type is shown in Fig. 4(a) and its diffraction pattern in Fig. 4(b). The electron micrograph shows bend contours and other diffraction contrast features but no grain or subgrain boundaries. The diffraction contrast features are thought to be largely due to the stresses caused by the expansion and contraction of the nitrocellulose protective layer during the tellurium dissolution. This layer absorbed the solvent and subsequently contracted when the film was dried after the final washing. Without the protective layer, however, the films shredded due to cracking in the (001) direction and were impossible to examine. The diffraction pattern shows no evidence of departure from monocrystallinity but does show forbidden reflections in the (0001) and (0002) positions. These are probably due to multiple diffraction events and are discussed in a previous paper.[6]

The influence of substrate temperature on the structure of the films is illustrated by the electron diffraction patterns shown in Fig. 5. At 40°C the films were uniform in thickness and, as far as could be detected, entirely amorphous. At 60°C, however, the structure was not reproducible, and varied from an amorphous droplet structure indicating some limited mobility of the atoms on the substrate to the complex ring pattern shown in Fig. 5(b). This latter structure appears to be a mixture of trigonal selenium and some other phase, possibly α-monoclinic. At 80°C the tendency towards single-crystal growth with the trigonal $(10\bar{1}0)$ [001] Se/$(10\bar{1}0)$ [001] Te orientation is obvious although some $(11\bar{2}0)$ [001] orientation is present. From 95° to

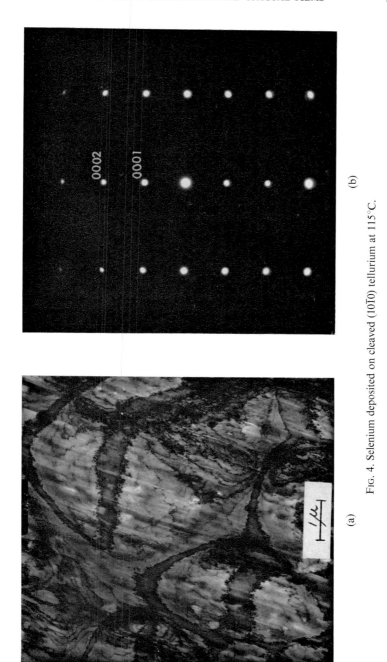

Fig. 4. Selenium deposited on cleaved (10$\bar{1}$0) tellurium at 115°C.

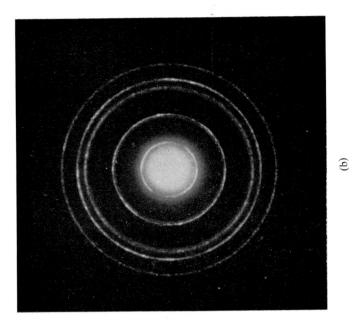

(b)

(a)

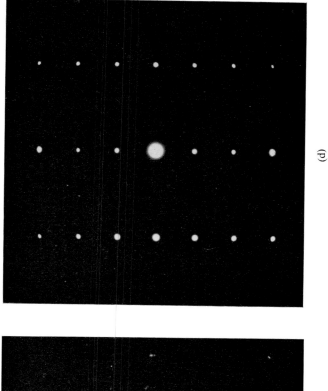

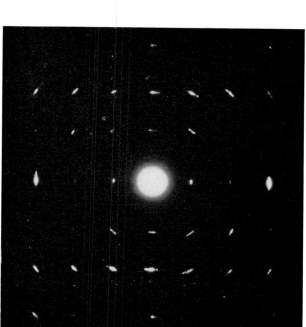

FIG. 5. The influence of substrate temperature on the structure of selenium films on (10$\bar{1}$0) tellurium: (a) at 40°C; (b) at 60°C; (c) at 80°C; (d) at 120°C.

135°C only the (10$\bar{1}$0) [001] Se orientation was observed and good single crystal films were formed as indicated by the presence of Kikuchi bands in diffraction patterns from the thicker films. These films were relatively two-dimensional in the growth, becoming continuous at about 500 Å. Lattice constants were determined by comparison with those from an evaporated sodium chloride film and were found to be in good agreement with the published values for bulk selenium. No indication of the presence of tellurium in these films was observed either in the electron-microscope images

FIG. 6. Dislocations in a (10$\bar{1}$0) plane selenium film.

or in the calculated lattice constants. This was verified for very thin (50 Å mean thickness) films where any diffusion of the tellurium would be obvious. Diffusion of copper, when copper specimen support grids were used, to give the stable copper selenide $Cu_{1.8}Se$ was, however, observed. This selenide has an identical structure and lattice constant with that reported by Andrievskii et al.[7, 8] for a β-cubic modification of selenium found in a thin film study.

The structures of films deposited on well-annealed polished substrates were very similar to those on cleaved substrates. Polished substrates were always used for the preparation of the thicker films (> 2500 Å) as cleavage steps in the substrate produced weak points and cracking in the films. Good single-crystal films up to 10 microns in thickness were produced with deposition rates in the order of 0.1 micron/min. Any greater deposition rate led to a falling off in perfection. Cracking due to differences in the thermal expansion coefficients of selenium and tellurium often occurred in the very thick films. Slow cooling was used to help overcome this problem.

The dislocation structure of the single crystal films was not studied in any detail; however, two distinct types of defects appeared to be present. These were line dislocations inclined to the ($10\bar{1}0$) film plane and dislocation loops, marked A and B respectively in Fig. 6. Both defects were mobile when illuminated by the electron beam and the loops in particular moved rapidly in the $\langle 001 \rangle$ direction. The motion of the lines, although generally in the $\langle 001 \rangle$ direction, was more complex.

Annealing of amorphous films deposited on the substrates at 40 °C either on, or after stripping from the substrate, resulted in the growth of small crystals with a common orientation. Annealing on the substrate at 90 °C gave films with single-crystal orientation and a grain size of about 500 Å. Annealing by electron bombardment in the electron microscope after stripping gave the structure shown in Fig. 7. Here individual single crystals can be seen growing in the amorphous matrix. It is obvious from the diffraction pattern that a considerable degree of order exists in the growth and that even when film deposition takes place at 40 °C there is ordering at the substrate overgrowth interface to provide the necessary nuclei. Selenium melted on the tellurium substrate and allowed to cool slowly also crystallized to give a single crystal orientation.

(b) (0001) *Tellurium Substrate*

Tellurium does not cleave in this plane and therefore selenium could only be deposited on polished and annealed surfaces. Films were prepared at substrate temperatures from 40° to 140 °C and epitaxial growth was observed from 60° to 135 °C. Excellent single-crystal growth took place between 100° and 135 °C and a typical example of this is shown in Fig. 8. The electron micrograph shows a structure which is more three-dimensional than that found on the ($10\bar{1}0$) face. Initially the deposit consisted of isolated crystallites which eventually joined by the filling in of the channels between them at a thickness of about 900 Å. Removal of the thin films from the substrates without damaging them was very difficult and therefore no conclusions could be drawn from the diffraction contrast contours present in the image.

The relation between substrate temperature and film structure is shown in Fig. 9. As with films on the ($10\bar{1}0$) face the structure at 40 °C was apparently completely amorphous. At 60 °C, however, the films began to grow epitaxially and there is obvious orientation shown in the diffraction pattern even though the spots lie on faint continuous rings. Trigonal $10\bar{1}1$, $10\bar{1}2$, $20\bar{2}1$, $11\bar{2}2$ rings indicating considerable polycrystallinity are also present. At 80 °C the

Fig. 7. Selenium deposited on (10$\bar{1}$0) tellurium at 40°C crystallized in electron beam.

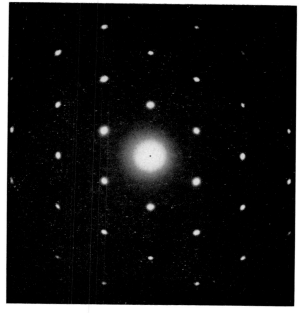

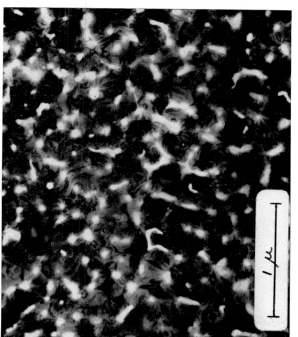

FIG. 8. Selenium deposited on polished and annealed (0001) tellurium at 130°C.

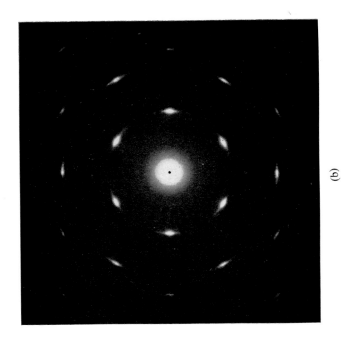
(b)

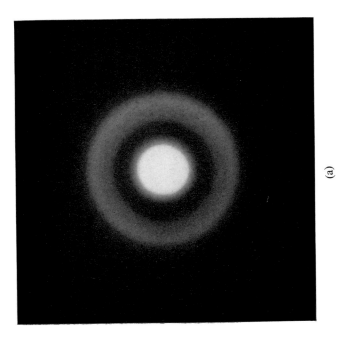
(a)

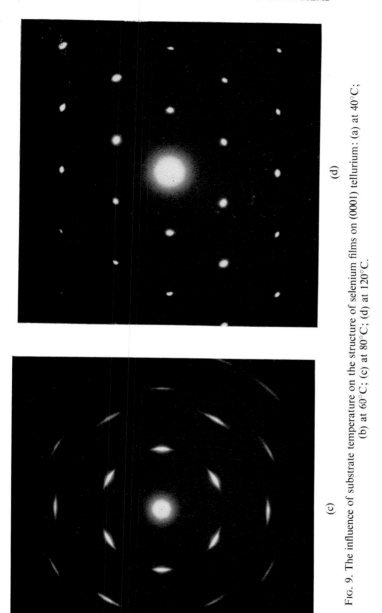

FIG. 9. The influence of substrate temperature on the structure of selenium films on (0001) tellurium: (a) at 40°C; (b) at 60°C; (c) at 80°C; (d) at 120°C.

diffraction pattern is basically single crystal with ±3° arcing of the spots, but faint arcs with indices $20\bar{2}1$, $30\bar{3}1$ and $31\bar{3}1$ are also visible. The latter reflections indicate some slight polycrystallinity in the form of deviations of the c-axis away from the normal to the film plane. At 100°C and up to 135°C the diffraction patterns all indicated good-quality single-crystal growth, and no change in microstructure was observed which could not be accounted for by the perfection of the substrate polish. Above 135°C no selenium deposited permanently on the substrates. Films deposited at 40°C and crystallized by electron bombardment in the electron microscope after removal from the

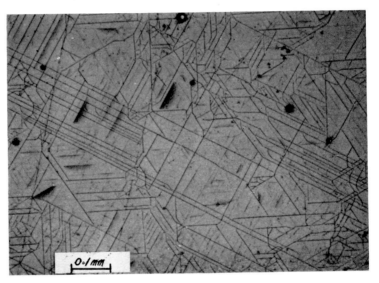

FIG. 10. Crack pattern in 2.5-micron thick (0001) selenium film.

substrate showed a considerable degree of orientation. This again shows as with the ($10\bar{1}0$) films that there must be a large degree of ordering at the substrate film interface even at 40°C.

As with the ($10\bar{1}0$) substrates, good-quality single-crystal films up to 10 microns in thickness were deposited on (0001) substrates at temperatures in the order of 100°C. These layers were particularly fragile and were subject to cracking on the substrate if the substrate temperature was lowered too rapidly after deposition. Cracking apparently occurred in the ($10\bar{1}0$) and ($11\bar{2}0$) planes as shown by the optical micrograph of a 2.5 micron thick film illustrated in Fig. 10. Line dislocations, in this case apparently parallel to the film plane, were again observed in the thinner films, but point defects (dislocation loops) were not seen.

(c) (0001) Bi_2Te_3 Substrate

Bismuth telluride has a layer structure with Van der Waals bonding in the (0001) plane. Smooth clean (0001) substrates could therefore be prepared by simply stripping a layer from the crystal with adhesive tape. Selenium was deposited on this surface at 100 °C to give films with thicknesses from 200 to 2000 Å thick. The structure of a typical film, in this case with a mean thickness 700 Å, is shown in Fig. 11. The orientation was not simple as was that on the tellurium substrates but a complex although precise mixture. It comprised one orientation of the (0001) plane parallel to the substrate and two triple orientations of the ($10\bar{1}0$) plane. Each of the two triple orientations consisted of three orientations of the selenium ($10\bar{1}0$) plane with their c-axes at 120°. The two triple orientations were superimposed and rotated through 25° with respect to one another. Using selected area diffraction it was found that individual islands (crystallites) of selenium had one of the orientations described. Figure 12(a) shows a pattern obtained from an island with the (0001) orientation and Fig. 12(b) shows the pattern from two adjacent islands with the ($10\bar{1}0$) orientation at 120°.

It seemed possible that by varying the film deposition parameters the selenium might be induced to take up one of the above orientations preferentially. Selenium was therefore deposited at 40 °C and then annealed at 90 °C for 1 hr on the substrate. The structure of this film was found to be similar to that of films deposited at 100 °C but with a smaller contribution from the (0001) orientation.

DISCUSSION

Although the factors controlling the overgrowth of deposited film material on a substrate lattice are not well understood, it is obvious that the interaction of the periodic field of the substrate with the depositing atoms will be very important. The understanding of this interaction is complicated by the possible rearrangement of atoms at the free surface of a crystal lattice to give a lower free energy configuration, and by the adsorption of gas or other impurity atoms on that surface. The latter complication is particularly evident in the case of films prepared under a vacuum in the order of 10^{-5} torr. Notwithstanding the above reservations the orientation of selenium films on the ionic alkali halides[5] does seem to be influenced by the register of substrate and overgrowth lattices. In the case of selenium on ($10\bar{1}0$) and (0001) tellurium, the selenium lattice is in parallel orientation to the tellurium (Fig. 13) and the lattice symmetry and bonding character are maintained across the substrate-overgrowth boundary.

With the (0001) Bi_2Te_3 substrate the structure presented to the arriving selenium overgrowth molecules is very interesting. The substrate face is

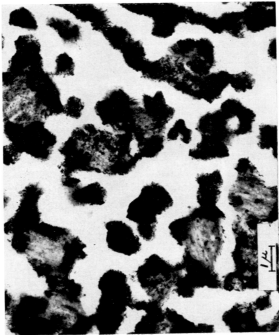

FIG. 11. Selenium on (0001) Bi$_2$Te$_3$.

THE GROWTH OF SELENIUM SINGLE-CRYSTAL FILMS 151

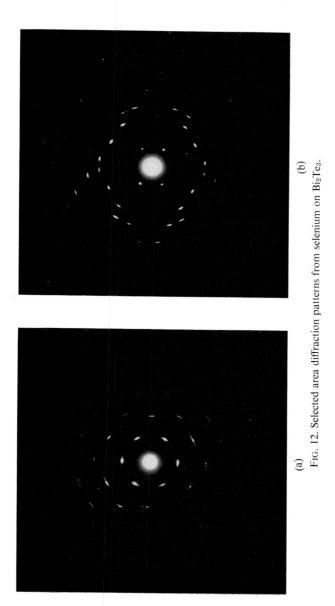

FIG. 12. Selected area diffraction patterns from selenium on Bi_2Te_3.

composed only of an hexagonal array of tellurium atoms with an interatomic spacing of 4.38 Å, compared to 4.49 Å for the (0001) face of tellurium itself. This is very close to the a_o spacing for selenium (4.36 Å) and would appear to be an ideal substrate for growth of the selenium (0001) face. The films deposited on Bi_2Te_3 at 100°C, however, show a mixture of orientation. It appears therefore that in the growth of selenium films in the (0001) plane the bonding across the substrate-overgrowth interface is very important.

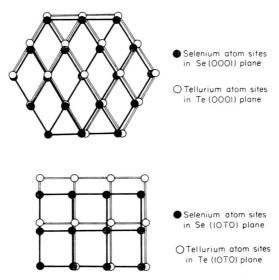

FIG. 13. Misfit diagram for selenium on (0001) and (10$\bar{1}$0) tellurium.

This seems reasonable in view of the fact that the selenium chains are normal to the substrate surface in this orientation and only bonded to the substrate by one atom. It is therefore much easier to grow single-crystal (10$\bar{1}$0) films than it is to grow (0001) plane films and in fact crystalline selenium films deposited on non-orienting substrates invariably have the c-axis parallel to the substrate.

The defect structure in the single crystal films was not studied in any detail. The mobility of these defects as observed in the electron microscope indicates, however, that the physical properties of selenium crystals should be very sensitive to stress and thermal history.

In electron diffraction patterns from selenium, reflections at "d" spacings, which do not correspond to the monoclinic or hexagonal modifications of selenium, were often observed. When examined in detail these were always found to be due to double diffraction effects[6] or to the diffusion of impurities into the film. No evidence of any modification of selenium other than the known monoclinic or hexagonal phases was ever found even when the amorphous phase was crystallized by heating or electron bombardment.

CONCLUSIONS

Large-area crystal films of selenium have been grown in both the ($10\bar{1}0$) and (0001) planes by epitaxial growth onto ($10\bar{1}0$) and (0001) tellurium. The quality and the purity of the films appear to be good up to a thickness of 10 microns and no evidence of the diffusion of tellurium into the film was observed.

Films have been removed from their substrates and optical measurements made at the absorption edge and on the short wavelength side of the edge. Preliminary data appear to be similar to those calculated from reflectivity measurements by Tutihasi and Chen.[9]

ACKNOWLEDGMENT

The authors would like to acknowledge the support of Canadian Copper Refiners Limited and the Directorate of Industrial Research of the Defence Research Board of Canada, and the assistance of P. T. Chiang and R. Dufresne who grew the tellurium crystals.

REFERENCES

1. KEEZER, R. C. and BAILEY, M. W., *Mat. Res. Bull.* **2**, 185 (1967).
2. KEEZER, R. C., WOOD, C. and MOODY, J. W., *Proc. Int. Conf. on Crystal Growth*, Pergamon, New York, 1967.
3. HARRISON, D. E., *J. Appl. Phys.* **36**, 1680 (1965).
4. BROWN, F. C., *Phys. Rev.* **4**, 85 (1914).
5. GRIFFITHS, C. H., *Recent Advances in Selenium Physics*, Pergamon, London, 1965.
6. GRIFFITHS, C. H. and SANG, H., *Mat. Res. Bull.* **2**, 515 (1967).
7. ANDRIEVSKII, A. I., NABITOVICH, I. D. and KRIPYAKEVICH, P. I., *Doklady Akad. Nauk SSR*, **124**, 321 (1959).
8. ANDRIEVSKII, A. I. and NABITOVICH, I. D., *Krystallografiya*, **5**, 465 (1960).
9. TUTIHASI, S. and CHEN, I., *Phys. Rev.* **158**, 623 (1967).

DISCUSSION

RICCIUS: Could you estimate roughly the electrical resistivity of the evaporated selenium films?

GRIFFITHS: In the order of 10^6 ohm-cm.

RICCIUS: What was the size of the area of the vacuum deposited selenium films after being stripped from the substrates? How thick was the thinnest film prepared by this method?

GRIFFITHS: The area of the films was restricted only by the size of the substrates available. In the case of the (0001) plane films, this was an area approximately 1.5 cm in diameter and in the case of the ($10\bar{1}0$) plane films an area approximately 1.5×2 cm. The thinnest films prepared were approximately 50 Å thick but these, of course, were not continuous.

HARRISON: Did the dislocation lines intersect the $(10\bar{1}0)$ surface, and did you find any evidence for dislocation pile-ups in the $\langle 10\bar{1}0 \rangle$ in the (0001) section?

GRIFFITHS: Yes, the dislocations did intersect the $(10\bar{1}0)$ surface in the $(10\bar{1}0)$ plane films but there was no evidence of pile-ups in the $\langle 10\bar{1}0 \rangle$ direction in the (0001) plane films.

WOOD: What was the degree of perfection, and how perfect were your thickest films compared with other methods of growth?

GRIFFITHS: The films up to about 2500 Å in thickness which could be examined in the electron microscope were very good single crystals with no evidence of grain boundaries or twinning. The diffraction patterns usually showed Kikuchi bands which are indicative of a high degree of perfection.

The perfection of the thicker films depended markedly on the deposition rate but at a deposition rate of one-tenth of a micron per minute films up to 10 microns thick gave very good Laue patterns. Hetero-epitaxial growth does, however, normally give a high dislocation density, but no measurements of this type have been made. Optically the films appear to be a little less perfect than the crystals grown from impurity doped melts but more perfect than any other crystals examined. No other method of growth, of course, allows the preparation of single crystals in this thickness range.

SOLUBILITY AND GROWTH OF TRIGONAL SELENIUM FROM AQUEOUS SULFIDE SOLUTIONS

E. D. KOLB

Bell Telephone Laboratories, Incorporated, Murray Hill, New Jersey

INTRODUCTION

Single crystals of trigonal (also called hexagonal) selenium are of interest for use as optical modulators[1-3] and parametric oscillators and amplifiers in the 10.6 μ region. The growth of single crystals having a high degree of perfection from the melt is hampered by the high viscosity of molten selenium. The liquid is composed of a mixture of selenium rings and long polymer-like selenium chains.[4] Keezer[5] used selenium melts doped with halogens or thallium to lower the viscosity for crystal growth by the Czochralski technique. Harrison[6] used argon at high pressure over the melt to increase the melting point and in this way decrease the viscosity. Impurities incorporated into crystals grown by the doping technique and the low degree of perfection of crystals grown at high pressure render the selenium unsatisfactory for modulator and amplifier studies. Stubb[7] reported the growth of small selenium crystals by the melt temperature differential technique and Eckart,[8] by vapor deposition.

The solubility of stibnite (Sb_2S_3) in Na_2S has been discussed by Arnston et al.[9] In a similar reaction I. Pouget[10] reported that selenium dissolved in alkali sulfides but not whether selenium was the stable solid phase.

Selenium crystallization from sodium sulfide solutions was observed in the author's laboratory during string-sawing experiments with selenium crystals grown by the author from thallium-doped melts. Amorphous selenium† had been added to dilute solutions of Na_2S in an attempt to reduce the string saw cutting rate and to obtain smoother cuts. Small metallic-appearing crystals were observed at the liquid container interface after the solution had been standing in a closed vessel for 2–3 days. The crystals gave an X-ray powder diffraction pattern for selenium. A Weissenberg photograph of one of the larger crystals proved that trigonal selenium had been grown.

† Amorphous selenium 99.999 per cent from American Smelting and Refinery Company, South Plainfield, N.J.

An attempt was made to obtain larger selenium crystals by growing onto a seed crystal immersed in 1.1 N Na₂S solution to which selenium had been added to a concentration approximating that of the string-sawing experiments. On slow evaporation, the surface of the liquid shortly became covered with a layer of selenium crystals; the largest crystals formed at the liquid–container interface.

Sodium sulfide solutions turn opaque when they become saturated with selenium. Consequently, dissolution of the seed crystal went unnoted until the seed suspension was withdrawn from the solution at the end of 3 days time. The solution in the vicinity of the seed apparently had not become saturated with selenium. The reason for this is possibly associated with the observed formation on the liquid surface of a selenium layer which retarded further evaporation of the solution.

The foregoing results suggested that slow cooling of the saturated solution rather than evaporation of solvent might be a better method of crystal growth. For this purpose it was necessary to determine the solubility of selenium as a function of temperature and Na₂S concentration.

The growth of monoclinic selenium as well as solubility data of selenium in carbon disulfide and various halogenated hydrocarbons has been reported by Moody and Hines.[11] They observed the formation of hexagonal selenium, but the solubilities reported were very low and so are not promising for the growth of hexagonal selenium crystals of substantial size.

SOLUBILITY METHOD

Initially, 50-ml stoppered glass flasks containing accurately weighed amounts of selenium in either 1.5 N or 2.5 N Na₂S, and a small magnetic spin bar were suspended in a 1500-ml beaker containing Nujol and a large magnetic spin bar. The observation of glass attack in preliminary experiments led to subsequent use of polypropylene flasks and stoppers. The large beaker was placed on a magnetic stirrer in such a way that the small magnet coupled to the large one provided stirring of both the Na₂S solution and the Nujol. The Nujol bath was heated externally by Briskeat tapes† properly insulated. The power was provided through a control relay, a Variac and a shunt resistor and was controlled by a thermo-regulator which was capable of controlling the bath temperature to $\pm 0.3\,°C$ up to $80\,°C$.

Solubility runs of 48 and 72 hr duration established that equilibrium had been reached by 72 hr. All solubility data reported in this paper were obtained for equilibration periods in excess of 72 hr. Following equilibration, the flask was removed quickly and its contents were emptied into a fine fritted glass filter positioned on a filter flask under vacuum. The solubility flask was rinsed

† Briscoe Mfg. Company, Columbus, Ohio.

repeatedly into the fritted filter and examined for any recrystallized selenium to insure complete recovery. Before weighing, the residue and fritted filter were rinsed with methanol and oven dried at 50°C.

SOLUBILITY RESULTS

Figure 1 is a plot of weight per cent of selenium in sodium sulfide solution at 1.5 and 2.5 N Na_2S from 26° to 80°C. There is a positive temperature coefficient of solubility from 25° to 65°C for 2.5 N Na_2S with evidence of retrograde solubility between 65° and 80°C. The coefficient is also positive

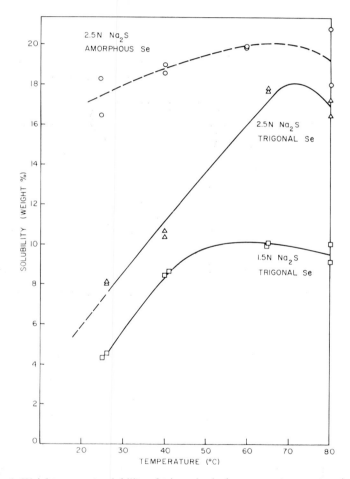

FIG. 1. Weight per cent solubility of trigonal selenium versus temperature in 1.5 and 2.5 N Na_2S and solubility of amorphous selenium in 2.5 N Na_2S.

from 25°C to 40°C for 1.5 N Na$_2$S, and levels off between 65°C and 80°C.

There is more scatter in the data at 80°C, and it is possible that at this temperature the dissolution reaction assumed to be sulfide complexing of the sort

$$Se + nS^= \rightleftharpoons SeS_n^{-2n}$$

is either no longer well represented by this equation or the value of n has changed.

The scatter may also be due to a slight leakage of H$_2$S at 80°C from the polypropylene flasks. It was difficult to maintain tight seals at this temperature and still be able to open the flasks rapidly for quenching and filtering. Similarly, the small amount of a red phase which appeared in the 80° temperature range may be monoclinic selenium, whose formation would introduce further scatter. The measured solubilities of amorphous selenium in 2.5 N Na$_2$S were higher at all temperatures (Fig. 1) and the results were more scattered than for trigonal selenium, as might be expected.

GROWTH METHOD

Following the preliminary crystal growth experiment a rotary crystallizer[†] was used to crystallize selenium from 1.5 N Na$_2$S by slow cooling from 50° to 25°C. Figure 2 is a drawing of the apparatus. The original glass tank was replaced with a 2000-ml flanged jar, jar clamps[‡] and a $\frac{3}{8}$-in thick plexiglass cover. A $\frac{1}{32}$-in. neoprene gasket was used on the cover of the flask and under the stirring motor to keep evaporation to a minimum. A three revolution per day clock motor together with no-slip pulleys[§] and belt drive were combined to drive the thermoregulator control magnet for the required slow cooling. The stirring shaft was modified by the addition of a second seed crystal support rod, and teflon paddles to increase stirring. (See Fig. 3a, at the left.)

A perforated platinum basket, secured to a $\frac{1}{8}$-in. stainless-steel rod, was arranged so as to be raised and lowered vertically into and out of the solution. The rod passed through an O-ring coupling fastened to the crystallizer plexiglass cover by a neoprene stopper. Selenium was added to the solution and the jar was sealed and held at 50°C for 36–48 hr, using a 10 rpm stirring speed. Following this, excess selenium was added to the platinum basket which was then lowered to the bottom of the crystallizer jar. Visual checks were made over a 24–36 hr period to determine whether solid selenium still remained in the basket. If the selenium test crystal dissolved, more selenium was added to the solution until solid selenium finally remained in the basket. Once saturation was established, cooling at the rate of 1°C per day was begun.

[†] Dependable Printed Circuit Corp., 365 Black Oak Ridge Road, Wayne, N.J.
[‡] Scientific Glass Apparatus Co., Inc., Bloomfield, N.J.
[§] PIC Design Corp., East Rockaway, New York 11518.

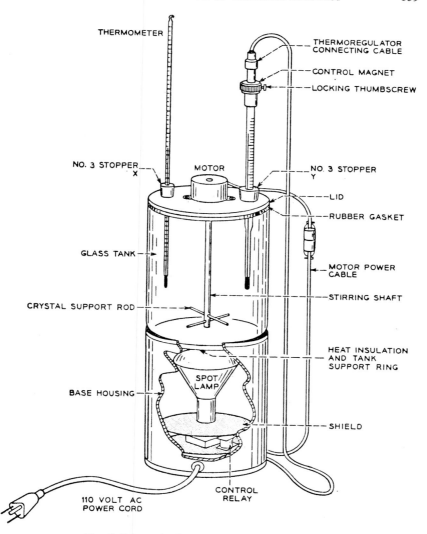

Fig. 2. Schematic diagram of rotary crystallizer.

GROWTH RESULTS

Figure 3a shows trigonal selenium crystallized on the teflon stirring paddles from the first run. The yield consisted of approximately 14 g of many fibrous selenium crystals 0.7–0.8 cm long by 0.3–0.5 mm thick and numerous rather well-formed hexagonal single crystals as large as 3–4 mm long by 1 mm thick (Fig. 3b). Major growth is along the c-axis and crystals perfectly

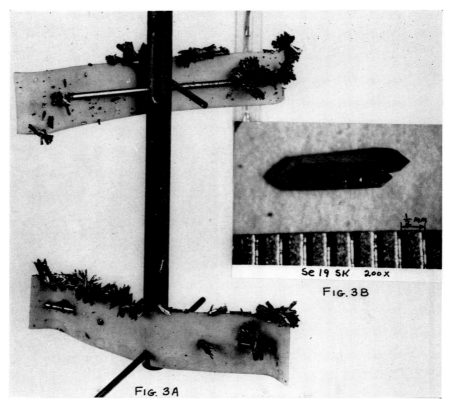

FIG. 3a. Trigonal selenium crystallized from Na_2S on teflon stirring paddles as fibrous needles and single crystals.

FIG. 3b. Enlarged view of single crystals of trigonal selenium.

hexagonal in cross-section had (100) prism faces and rhombohedral faces of the (313) form (indexing is in the hexagonal system). It was also determined by X-ray goniometry that crystals not perfectly hexagonal in cross-section contained both (100) and (110) faces parallel to the c-axis.

Emission spectroscopy revealed silicon to be present to the extent of <50 ppm rather than the <1 ppm of the starting selenium. This is due to attack of the Pyrex jar by the Na_2S. One preliminary wet chemical analysis for sulfur revealed it to be 170 ppm.

The next experiment was an attempt to grow onto seeds cut and cleaved from Czochralski grown crystals, in 1.5 N Na_2S by starting the cooling cycle at 48°C. Despite the fact that the platinum basket presaturation technique was used prior to seed introduction, the seeds dissolved. It is obvious from Fig. 1 that better control of the degree of supersaturation will be attained by growing from 2.5 N Na$_2$S. In the more concentrated solution, the slope of the solubility curve is steeper and positive to 65°C. This feature should make it considerably easier to retain seeds started at 50°C. Initial experiments indicate that complexing reactions for the growth of other materials warrant investigation.

ACKNOWLEDGMENT

The author wishes to acknowledge the contribution of R. L. Barns and J. L. Bernstein for the X-ray orientation data. Gratitude is expressed to W. C. Ellis for the initial selenium identification by Weissenberg photograph, to D. J. Nitti for powder X-ray identification, D. L. Nash for the analysis data, and to R. A. Laudise for helpful discussions concerning the chemical complexing aspect of this and future investigations.

REFERENCES

1. KAMINOW, I. P., *IEEE J. Quant. Electr.*, to be published.
2. TEICH, M. C. and KAPLAN, T., *IEEE J. Quant. Electr.* **2**, 702 (1966).
3. PATEL, C. K. N., private communication.
4. EISENBERG, A. and TOBOLSKY, A. V., *J. Polymer Sci.* **16**, 19 (1960).
5. KEEZER, R. C. et al., *Proc. Int. Conf. Crystal Growth*, June 1966, Pergamon, New York, 1967.
6. HARRISON, D. E., *J. Appl. Phys.* **36**, 1680 (1965).
7. STUBB, T., *Recent Advances in Selenium Physics*, Pergamon, London, 1965.
8. ECKART, F., *Recent Advances in Selenium Physics*, Pergamon, London, 1965.
9. ARNSTON, R. H., DICKSON, F. W. and TUNELL, G., *Science*, **153**, 1673 (1966).
10. POUGET, I., *Ann. Chim. Phys.* **18** (7), 505 (1899), reprinted in J. W. MELLOR, *A Comprehensive Treatise on Inorganic and Theoretical Chemistry*, Vol. X, Longmans Green London, 1930.
11. MOODY, J. W. and HINES, R. C., *Mat. Res. Bull.* **2**, 523 (1967).

THE MORPHOLOGY AND GROWTH OF TRIGONAL SELENIUM CRYSTALS

C. H. Griffiths† and Brian Fitton‡
Noranda Research Centre, Pointe Claire, Quebec, Canada

INTRODUCTION

The morphological and structural details of the crystallization of trigonal selenium are presented here and are discussed using some of the ideas now current in descriptions of the crystallization of the organic high polymers. Such an approach was suggested by the postulated polymeric nature of selenium melts.[1]

The crystallization of supercooled melts of linear polymers has been reviewed extensively.[2] Such crystallization shares with selenium the unusual feature that the crystalline regions develop in the form of spherulites. A consideration of the energetics leading to such a preferred type of growth has been made by Fullman.[3] Studies[4] of the internal structure of these organic polymer spherulites have shown clearly that they consist of a system of thin plate-like lamellae of almost constant width which radiate from a common center and branch non-crystallographically. The molecular chains were found almost invariably to be oriented normal to the radius and hence normal to the growth direction. The widths of these lamellae are generally very much less than the known chain lengths, leading to the now established concept of a lamella growth mechanism involving repetitive 180° chain folding to form the edges of the lamellae. The details of such folding have been examined by Lindenmeyer[5] and Peterlin.[6] The fold period appears to be independent of molecular weight[7] but increases with the crystallization temperature. The concentric-ringed structures which are a notable feature in sectioned spherulites of both selenium and the organic polymers have been shown, for the latter, to be the result of the lamellae twisting co-operatively along the radial direction during growth so that there is an in-phase progressive rotation of the molecular chain axes along the radius.

Consideration of the previous work on selenium crystal morphology[8] indicated similarities between selenium and organic polymer spherulites. A detailed structural examination of selenium spherulites was therefore undertaken to confirm this. The investigation was extended to a study of selenium single crystals grown from the melt.

† Present address: Research and Engineering Center, Xerox Corporation, Webster, N.Y.
‡ Present address: European Space Research Laboratory, Noordwijkerhout, Holland.

EXPERIMENTAL

Crystallization was studied in thin film and bulk amorphous selenium. The former allowed a detailed examination of structure and orientation in the electron microscope and the latter provided general confirmation of these results over a much wider temperature range.

Commercial high-purity selenium (99.99+% Se) supplied by Canadian Copper Refiners Limited was vacuum distilled to provide the basic material for these studies.

The films of selenium were prepared by first evaporating the selenium at 260°C onto nitrocellulose-coated microscope slides. These were subsequently heated to 80°, 90° or 100°C to initiate crystallization of the selenium. The nitrocellulose-supported films were removed from the glass by immersion in water and collected on copper grids for examination in the electron microscope.

For studies of bulk crystallization the selenium was sealed under vacuum in silica and heated at 480°C for 12 hr after which the ampoules were transferred to a furnace held at the appropriate crystallization temperature. The samples were subsequently fractured and two-stage carbon replicas of the fracture surface were taken for optical or electron-microscope examination. Carbon shadowing at 45° was generally used for the latter. Crystallographic information was obtained from electron-diffraction examination of crystal fragments attached to the surface replicas.

Small single crystals (1 mm long) were grown from a melt of vacuum distilled selenium held at 215°C. Large single crystals grown by the Bridgeman technique under a pressure of 5 kilobars and by Czochralski and traveling solvent methods from impurity-doped melts were obtained and examined. Fresh cleavage surfaces of these crystals were prepared by stripping with adhesive tape and were examined optically or electron optically using the above-mentioned replication process.

RESULTS

(a) *Spherulite Growth*

In the temperature range from 80°C, the lowest temperature studied, to 200°C, amorphous selenium crystallizes in the form of spherulites. These were predominantly of two types. The first (type I) was found in both thin film and bulk samples. The other (type II) was found when the spherulites developed in contact with certain foreign surfaces. Figure 1 is a typical example of a fractured section of a type I spherulite. In this case the crystallization temperature was 80°C. Both the radiating lamellae and the ringed structure are clearly defined.

Fig. 1. Fracture section of spherulite grown at 80°C in sample of vacuum distilled high purity selenium.

No specific study of the nucleation process in spherulite growth was attempted. However, sheaf-like nuclei were observed frequently in the bulk samples over the range 100–200°C and also in the thin films. Such nuclei have also been commonly observed in organic polymers.[9] The nucleus is a bunch of aligned single crystals which eventually branch from both of the ends. Progressive branching causes each end to fan out so that the whole structure looks rather like a wheatsheaf (Fig. 2).

Fig. 2. Development of a wheatsheaf spherulite in a bulk selenium sample at 188°C.

The orientation and microstructure of thin film spherulites was determined using electron microscopy and selected area electron diffraction. High-magnification examination was complicated by the rapid recrystallization and grain growth that resulted from the high electron-beam intensity. It was clear, however, that the type I spherulites grown in films from 400 to 1500 Å thick at 80–100°C were composed of radially oriented lamellar crystals in the same manner as those observed optically following crystallization at higher temperatures. In the sample crystallized at 100°C the lamellae were between 400 and 800 Å wide and as thick as the film itself (1000 Å). Figure 3(a) shows such a structure together with the frequently observed dark annular ring.

THE MORPHOLOGY OF TRIGONAL SELENIUM CRYSTALS 167

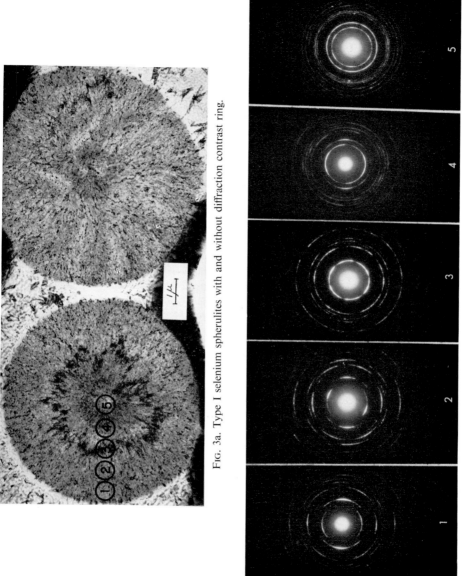

FIG. 3a. Type I selenium spherulites with and without diffraction contrast ring.

FIG. 3b. Electron-diffraction patterns showing change in orientation along the radius of type I selenium spherulites.

Fig. 4. Type II dendritic form of selenium spherulite grown in contact with soda glass at 180°C.

The orientation of the lamellae was determined using selected area diffraction. A number of spherulites with, and without, the dark annular ring were examined and in all cases the spherulites were found to be roughly centro-symmetric with the orientation parallel to the spherulite plane changing uniformly with the increasing radius. Figure 3(b) shows electron diffraction patterns from the areas indicated in Fig. 3(a). The orientation change is seen to be caused by a twist about the radial (210) axis of roughly 360° per 18 microns of radius. The dark ring was observable only in those spherulites

FIG. 5. Electron-diffraction pattern from arm of type II selenium spherulite.

in which the lamellae twisted co-operatively so that the progressive changes in orientation along all radii were in phase. The position of the ring was found to be coincident with the point where the (0001) plane was parallel to the spherulite plane. This ring then is an extinction contour in each of the separate crystals from planes with the [001] zone axis. Thus the twisting of the lamellae during growth, which is a feature also of organic polymer spherulites and which is anticipated from an examination of selenium spherulite fracture surfaces with crossed polarizers, is here confirmed.

Spherulites which developed in contact with silica containers were primarily of the same form and structure as those developed in the bulk of the sample. However, in contact with Pyrex and normal soda glass the spherulites grew at rates at least a factor of ten faster than in the bulk and with a quite different morphology. Figure 4 shows a well-developed example of this form,

Fig. 6. Carbon-shadowed replica of the surface of a type II selenium spherulite.

called type II, grown at 180°C. At lower temperatures the individual radiating arms tended to be of a finer form and to twist, but the basic almost dendritic branching from the center spine could still be resolved. Such structures were also found in the thin films examined with the electron microscope. Using selected area diffraction, the orientation radially out along the arms and its relationship to the orientation of the material between the arms, was investigated.

The diffraction pattern obtained was constant radially out along the arm and also from one spherulite to another. This pattern, shown in Fig. 5, was found to be a composite made up of three orientations of a single primary pattern each corresponding to an orientation of the hexagonal selenium $(11\bar{2}2)$ plane. The three contained [111] directions were respectively radial,

Fig. 7. Leaf-type selenium single crystals showing 90° twinning.

and 45° either side of the radius. The crystallographic orientations described above and the appearance of the observed diffraction contrast contours are consistent with the center of the arm having the [111] radial orientation and the sides of the arm having plus and minus 45° orientations produced by twinning on the ($\bar{1}$101) plane. Confirmation of this radial orientation for bulk form samples was obtained from diffraction patterns produced by fragments adhering to replicas.

Figure 6 is an electron micrograph from a carbon replica of the surface

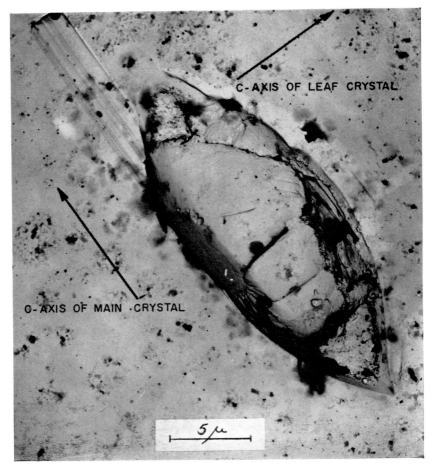

FIG. 8. Leaf crystal embedded in main selenium single crystal which had been grown by the traveling solvent method with thallium doping.

shown in Fig. 4. It is apparent that the detailed structure is more complex than indicated by an electron-microscope examination of those dendritic forms grown in thin films. The basic components are again lamellae. They are apparently directed parallel to the radial arms but inclined at a small angle to the interface with the foreign surface. Their broad faces are exposed at the junction with the smooth center depression. If it is assumed that the lamellae are edge on to the top surface with the molecules contained across the width of the broad faces, as in normal spherulite lamellae, then the c-axis direction can reasonably be accommodated in the direction suggested by the diffraction analysis discussed above. The general growth arrangement seen in these photographs of bulk samples is therefore consistent with the structural and orientational details derived from the thin film studies, and with the orientations obtained from diffraction by fragments of these radial arms found attached to replicas.

(b) *Single Crystal Growth*

Figure 7 shows some of the leaf-shaped single crystals commonly found in samples crystallized at above 210°C. The maximum size of leaf crystal so far observed was 1.9 mm long × 0.9 mm wide with a thickness (0.3 mm) equal to that of the sample. Usually the dimensions were limited by intergrowth with other crystalline material. Electron-diffraction patterns from fragments attached to replicas of the surfaces of such single crystals indicated that the c-axis is directed parallel to the minor axis of the "leaf" in agreement with Ehinger.[8]

Spiral pyramidal growths were frequently observed on the free-growing surfaces of these leaf crystals. Twinning was also commonly found in these

FIG. 9. Extended chain lamellae in high-pressure-grown selenium crystals.

leaf crystals to give a second leaf with the major axis at 90° to the first as shown in Fig. 9.

Single crystals grown elsewhere by the traveling solvent and Czochralski methods which have been examined in this study have generally shown the presence of these small leaf-like single crystals within the main body of the crystal (Fig. 8). They are seen to lie with their major axis parallel to the c-axis of the main crystal. Thus their c-axis is at right angles to the main crystal c-axis and also to the growth direction.

The selenium single crystal grown from a high purity melt at 5.3 kilobars pressure of argon showed large numbers of elongated voids, probably due to occluded argon[10] in an infrared transmission photograph. Figure 9 shows a cleavage face of this crystal, the c-axis lying parallel to the striations. Perpendicular to these striations numerous low-angle boundaries are to be seen. These boundaries frequently extend through the width of the ingot (approximately 1 cm) with a constant width of crystal contained within them. Such widths vary from 10^{-3} cm up to 0.6 cm.

DISCUSSION

The mean length of the selenium chain in the neighborhood of the melting point, as suggested by the work of Keezer and Bailey,[11] is about 20 microns. It is obvious, therefore, that the selenium chains must be folded in some way in order to be accommodated within the width of the lamellae observed in spherulites grown below 210°C. Such a fold might be realized by a rotation of the Se–Se bond by 120°. With a consequent local dilation of the lattice at this fold, the stress produced along both edges of the growing lamella will cause it to twist about its growth axis. Details of the mechanism providing for this type of cooperative twisting of the lamellae in organic polymers and the resulting annular ring structure have been published elsewhere.[12] The measured mean lamellar widths in bulk samples varied from 0.1 micron at 80°C to a maximum of about 16 microns at 210°C. The latter value was then essentially constant up to the melting point, and presumably corresponded to extended chain crystallization.

From the foregoing it is clear that the selenium lamellae are essentially the same in both growth habit and structure as those lamellae which develop in spherulites of the organic high polymers. It follows therefore that the observed features of crystal growth from the selenium melt are primarily a result of its polymeric nature and not a property of the individual atoms.

Since the conformation and twisting of these lamellae within a bulk grown selenium spherulite are identical to those in the majority of organic polymer spherulites studied,[13] it is anticipated that the kinetics of spherulite growth will also follow closely the theory derived for a general polymer.[14, 15] The results obtained so far on the crystallization kinetics tend to confirm this.

In considering the possibilities of single-crystal formation it is necessary to acknowledge that, in theory at least, the terminal atoms in the selenium chain are not chemically distinct groups, as is the case with the organic polymers. Consequently in selenium, chain-end to chain-end bonding might be anticipated during normal single-crystal growth. This would contrast with the highly defective structures which appear as a result of the inclusion of the stable chain end groups during organic polymer single-crystal growth.[16] Recent e.s.r. measurements,[17] however, show the chain ends as being terminated by impurities or spinless groupings and indicate that such bonding is unlikely in undoped selenium. Any attempt to grow selenium single crystals from the melt might therefore be expected to produce mainly extended chain or chain folded structures with a high defect concentration. If, however, the melting point is raised by increasing the pressure above the melt, the increased mobility of the chains would facilitate a fractionation process to produce a range of distinct extended-chain lamellae each of which is a crystal having a distinct small range of chain lengths. This seems to be a plausible explanation for the existence of the low-angle boundaries found in the high-pressure-grown selenium single crystal, particularly in view of similar results obtained with high-pressure-grown organic polymer crystals.[18] The validity of this fractionation process model has been proved by Anderson[19] and Pennings and Keil[20] using organic polymers of known molecular weight distribution.

The structures discussed above would seem to have important implications as far as the transport properties of crystalline selenium are concerned. Thus the physical origin of the barriers so often alluded to in discussions of the electrical properties of selenium may well find an explanation in terms of such arrays of chain ends or of chain folds as those discussed above. In addition the inclusion in low-temperature growth of large numbers of chain ends in random positions within the crystal leads to a very high defect concentration. The mobility of such defects has been shown[21] to be very sensitive to temperature. This temperature sensitivity will lead to hysteresis effects in the physical properties as the chain ends migrate to give a lower energy configuration.

The leaf-like single crystals of selenium grown in the bulk just below the melting point appear, from the structural point of view, to be simply enlarged lamellae which nucleate and grow independently. They are quite distinct from solution or melt grown organic polymer single crystals. However, these leaf crystals also differ from the selenium spherulite lamellae in that they grow in all directions simultaneously, i.e. they have no set width. This militates against the idea of a chain-folded structure. It seems, therefore, that they are extended-chain crystals which originate as a bundle of aligned chains. The bundle-like growth of lamellae was originally considered by Hoffman and Lauritzen[14] who concluded that in general such lamellae would probably not be capable of growth to a large size. Here, however, they appear to be stable entities. The equilibrium shape of a lamella attempting to grow by a parallel alignment of chain segments tends to be ellipsoidal due to a cumulative strain induced by the density difference of the crystalline and non-crystalline phases.

The presence of these leaf crystals in "single crystals" grown from doped melts may well be due to the build-up at the growth front of unmodified high-molecular-weight chains terminated by impurities. These could then condense to form the bundle-like lamella suggested above. The alignment of the major axis of the leaf along the c-axis of the main crystal and the major and minor axes within the plane of the melt interface indicates a growth contemporaneous with the surrounding layer of main crystal and nucleation on some structural feature of the underlying single-crystal surface. This orientation relationship is identical to that found previously in the case of leaf crystal grown from the pure melt at 215°C and illustrated in Fig. 7. It appears, therefore, that this observed orientation relationship is due to the same twinning mechanism.

CONCLUSIONS

Selenium has been shown to behave as a typical high polymer so far as spherulitic crystallization is concerned. At the moment there is some uncertainty concerning the structure of the melt, particularly in relation to the proportion of rings and chains.[22] Whatever the equilibrium nature of the melt, however, the morphology and growth kinetics of the trigonal phase indicate that the crystallization is governed by the presence of long-chain molecules at the interface of the growing crystal and the liquid. Selenium differs from the organic high polymers apparently in the formation of leaf-like single crystals which do not contain chain folds, but whose formation and shape may be explicable in terms of a bundle-like lamellar growth. The growth of large selenium single crystals of low defect concentration would seem to be unlikely in view of the termination of chains by impurities which reduce the possibilities for chain to chain bonding. Only by doping the melt with chain terminating impurities which can subsequently be rejected from the crystal at the growth face would large single crystal growth appear feasible. This apparently occurs in the case of the Czochralski and traveling solvent crystals grown from impurity-doped (thallium, etc.) melts. The alternative of growth from the vapour would appear the more favorable method for obtaining samples of pure single crystals.

Finally it is concluded from a study of presently available single crystals that chain ends and folds play an important part in determining gross crystal perfection and require serious consideration in the analysis of the transport properties of such selenium crystals.

ACKNOWLEDGMENTS

The authors would like to acknowledge the support of Canadian Copper Refiners Limited and the Directorate of Industrial Research of the Defence Research Board of Canada, and the technical assistance of F. Rosenblum and G. Borsohalmi.

REFERENCES

1. EISENBERG, A. and TOBOLSKY, A. V., *J. Polymer Sci.* **46**, 19 (1960).
2. HOFFMAN, J. D., *S.P.E. Trans.* **4** (4), 315 (1964).
3. FULLMAN, R. L., *Acta Met.* **5**, 638 (1957).
4. GEIL, P. H., *Polymer Single Crystals*, Interscience, New York, 1963.
5. LINDENMEYER, P. H., *Science*, **147**, 1256 (1965).
6. PETERLIN, A., *J. Appl. Phys.* **31**, 1934 (1960).
7. LINDENMEYER, P. H. and HOLLAND, V. F., *J. Appl. Phys.* **35**, 55 (1964).
8. EHINGER, H., *Zeit. fur Kristallographie*, **115**, 235 (1961).
9. KELLER, A. and WARING, J. R. S., *J. Polymer Sci.* **17**, 447 (1955).
10. HARRISON, I. D. and HARRISON, D. E., *The Physics of Selenium and Tellurium*, Pergamon, New York, 1969, p. 115.
11. KEEZER, R. C. and BAILEY, M. W., *Mat. Res. Bull.* **2**, 185 (1967).
12. PADDEN, F. J. and KEITH, H. D., *J. Polymer Sci.* **51**, 54 (1961).
13. KELLER, A., *J. Polymer Sci.* **17**, 291, 351 (1955).
14. HOFFMAN, J. D. and LAURITZEN, J. I., *J. Res. Natl. Bur. Stds.* **65A**, 297 (1961).
15. FRANK, F. C. and TOSI, M. P., *Proc. Roy. Soc. (London)*, A **263**, 323 (1961).
16. PREDECKI, P. and STATTON, W. O., *J. Appl. Phys.* **37**, 4053 (1966).
17. ABKOWITZ, M., *J. Chem. Phys.* **46**, 4537 (1967).
18. GEIL, P. H., ANDERSON, F. R., WUNDERLICH, B. and ARAKAWA, T., *J. Polymer Sci.* **A2**, 3707 (1964).
19. ANDERSON, F. R., *J. Appl. Phys.* **35**, 65 (1964).
20. PENNINGS, A. J. and KEIL, A. M., *Kolloid Z.* **205**, 160 (1965).
21. HOLLAND, V. F., *J. Appl. Phys.* **35**, 3235 (1964).
22. AVERBACK, B., *Second Conference on the Characterization of Materials*, November 1967, Rochester, New York, U.S.A.

THE GROWTH OF α- AND β-RED MONOCLINIC SELENIUM CRYSTALS AND AN INVESTIGATION OF SOME OF THEIR PHYSICAL PROPERTIES

G. B. ABDULLAYEV, Y. G. ASADOV and K. P. MAMEDOV

Institute of Physics, Academy of Sciences of the Azerbaidzhan, Baku, U.S.S.R.

INTRODUCTION

The extensive use of selenium in semiconductor devices has resulted in a detailed investigation of the temperature dependence of structural changes.

It is known that selenium exists in the amorphous, hexagonal (or trigonal) and α- and β-monoclinic forms. Growth conditions and the crystallization mechanism of the hexagonal modification of selenium from the vapor phase, the melt, and the amorphous state are poorly understood, in spite of numerous investigations.[1-3] This situation is caused mainly by the fact that amorphous and liquid selenium represent an equilibrium mixture of at least two molecular modifications, a monomer, Se_8, and a polymer, Se_x, which have different binding (Se_8 is a ring monomer, Se_x is a chain-like polymer). The ring monomer may act as a source of the chain polymer and vice versa ($nSe_8 \rightarrow mSe_x$ or $Se_x \rightarrow Se_{x-n8} + nSe_8$).

Depending on thermodynamic conditions (temperature and pressure), crystallization (i.e. polymerization) from the amorphous and liquid phases takes place only to the hexagonal modification of selenium which is stable up to the melting point, 218°C. This fact shows that the ring molecules of Se_8 decay into the short chains with radical-like ends. All the physical properties of hexagonal samples depend on the percentage nSe_8 and mSe_x and their orientations.

From this point of view and because of the limited investigation of these substances, it is interesting to mention:

1. the growth of monocrystals of the monoclinic modification of selenium (which are composed only of molecules of eight-atom closed rings);

2. the investigation of the monocrystal of monoclinic selenium, the conversion of the monocrystal of monoclinic selenium into hexagonal form, and

3. the investigation of the physical properties of monoclinic selenium.

The monoclinic modification, in comparison with the amorphous and hexagonal form of selenium, has been poorly studied.[4-6] Evidently this situation is connected with the difficulties of obtaining monocrystals of monoclinic selenium of such dimensions as to permit investigations of the physical properties.

THE GROWTH OF SINGLE CRYSTALS OF MONOCLINIC SELENIUM

Monocrystals of the α- and β-monoclinic modifications of selenium as distinguished from the hexagonal modification are grown from a solution of amorphous selenium in carbon disulfide.[4, 7] To obtain a monocrystal of monoclinic selenium, as was done in the work of Prosser,[4] a flask with the reverse refrigerator was used in the present study. The apparatus consists of a furnace designed for heating a solution and thus increasing the solubility of amorphous selenium in carbon disulfide, of a flask 1.5 liters in volume with a reverse refrigerator and a cooler at the end of which monocrystals are grown.

The original material used to obtain a solution was selenium of B-5 type. To obtain amorphous selenium (the carbon disulfide dissolved only the amor-

FIG. 1. Photograph of monoclinic selenium monocrystal (30×).

phous selenium) the fused selenium was poured out into liquid nitrogen and from the resulting dark-red amorphous selenium, a fine powder was prepared.

Selenium and its solution in carbon disulfide were heated in the flask of the furnace at a constant temperature of 30 °C (b.p. $CS_2 = 46$ °C). The cooler at the end of which the crystals are grown had a temperature of 15 °C. Due to the temperature gradient between the heater and the cooler, the solution is constantly circulating by convection. The solution becomes saturated at a temperature of 30 °C by flowing past solid selenium at the bottom of the flask. Then it passes through the cooler region at lower temperature (15 °C) where the growing crystal is situated. Here the solution becomes supersaturated and thus conditions for growth are created. The material used for the growth of the crystal is replenished at the expense of selenium in the hot section, and the cycle is repeated.

Using the above procedure monoclinic monocrystals of selenium, mainly of $3 \times 3 \times 2$ mm³ dimensions, were obtained for the investigation of physical properties (Fig. 1).

For X-ray and microscopic investigations thin platelet crystals of α- and β-monoclinic selenium of $1 \times 1 \times 0.3$ mm³ dimensions were obtained (Fig. 2).

From the X-ray analysis of these crystals the parameters of the unit cells of the α- and β-monoclinic selenium and of the hexagonal selenium have been determined (after the full conversion of the α- and β-monoclinic forms into the hexagonal form). The values coincide exactly with the literature data[8, 11] given in Table 1.

TABLE 1. *Crystallographic Data for Monoclinic and Hexagonal Selenium*

Crystalline modification	Parameters of the unit cell							Spatial group
	a_0, Å	b_0, Å	c_0, Å	angles	V, Å	Z	ρ g/cm³	
α-Monoclinic	9.05	9.05	11.61	90°46	982.85	32	4.46	$C_{2h}^5 (P2_1/n)$
β-Monoclinic	12.85	8.07	9.31	93°8	963.99	32	4.50	$C_{2h}^5 (P2_1/a)$
Hexagonal	4.35		4.95		93.86	3	4.82	$D_3^4 (C3_1 21)$

In his structural determinations Burbank[9, 10] concluded that the eight-atom closed rings of the β-monoclinic form are deformed and one of the Se–Se distances in the molecule is larger by 0.73 Å than the corresponding distance in the α-form. On this basis Burbank concluded that the monocrystal of β-monoclinic selenium can be converted directly into the monocrystal of hexagonal selenium.

Crystals of α-monoclinic selenium are also subject to conversion into the hexagonal form at room temperature but at a lower rate than in the case of the β-monoclinic crystals.

Burbank's hypothesis about the transition of the α- and β-monoclinic

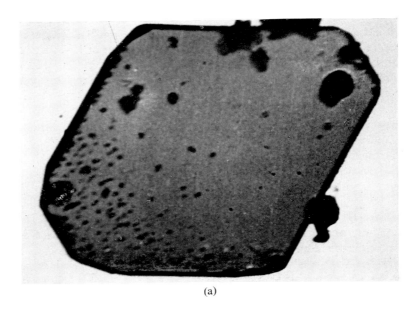

(a)

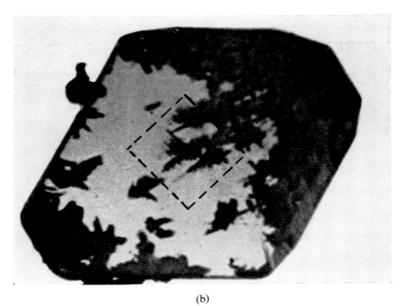

(b)

FIG. 2(a)–(c). For legend see p. 183.

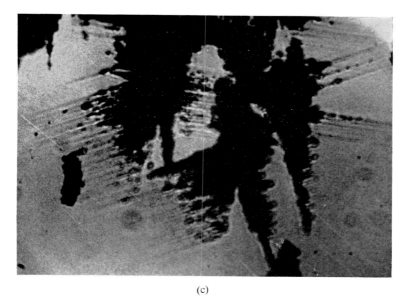

(c)

Fig. 2. Photomicrograph of the conversion of monoclinic selenium monocrystal into hexagonal selenium: (a) monocrystal of monoclinic selenium having the shape of plates; (b) the dendritic growth of hexagonal selenium inside the monoclinic monocrystal of selenium (60×); (c) the enlarged form of the part "b" separated by the dotted lines (1350×).

selenium into the monocrystal of hexagonal selenium at room temperature is erroneous as will be shown by experimental evidence set forth below.

Marsh et al.[11] showed that the configurations of the α- and β-monoclinic selenium molecule have a strong resemblance. In Fig. 3 the molecules of α- and β-monoclinic selenium are given according to Burbank's data and the molecule of β-monoclinic selenium according to the data of Marsh et al. The difference of the parameters of the unit cell of α- and β-monoclinic selenium is explained only by the different packing of the molecule.

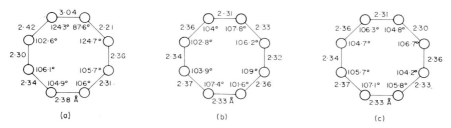

Fig. 3. Molecules of α- and β-monoclinic selenium: (a) the molecule of β-Se (Burbank); (b) the molecule of α-Se (Burbank); (c) the molecule of β-Se (Marsh et al.).

THE GROWTH OF THE HEXAGONAL MODIFICATION OF SELENIUM FROM MONOCLINIC SELENIUM

Until now there is only an hypothesis on the mechanism of growth of hexagonal selenium crystals from monoclinic selenium monocrystals.[9, 11]

The explanation of the mechanism of structural change in a solid requires a detailed investigation of the nucleation and growth velocity and of crystallographic bonding between the growing and the matrix crystals. Such an investigation may be carried out by microscopic and X-ray methods.

For the microscopic investigation a polarizing microscope is fitted with a heating table and a photocamera for microphotography. The sample rests on the objective table between heaters which provide a constant temperature in the space where the sample is placed thus preventing temperature gradients in the sample. The thermoregulator controls the temperature of the sample to $\pm 0.2\,°C$.

Optical observations and Laue photographs obtained at 120°C show that α-red selenium is not converted into β-red selenium as has been claimed.[12] The monocrystals of α- and β-red monoclinic selenium (without an interconversion depending on the temperature and the time) are converted into a grey hexagonal modification.

The conversion of $Se_{monoclinic} \rightarrow Se_{hexagonal}$ takes place with nucleation

(a)

FIG. 4(a)–(c). For legend see p. 185.

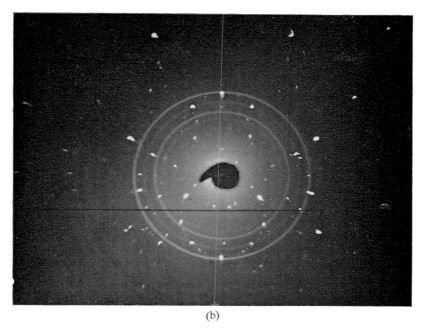

(b)

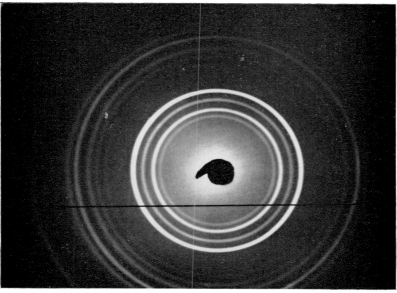

(c)

FIG. 4. (a) Laue pattern of the monocrystal of monoclinic selenium. (b) Laue pattern of monoclinic selenium partially converted into the hexagonal form. (c) Laue pattern taken after full conversion of the monoclinic monocrystal of selenium into the hexagonal form.

and immediate growth of hexagonal selenium.[13] The nuclei of hexagonal selenium inside the monocrystal of monoclinic selenium grow in the form of dendrites (Figs. 2 b and c). The Laue photograph of the monocrystal of monoclinic selenium (a), the Laue photograph of the monoclinic monocrystal of selenium partly converted into the hexagonal form (b), and the Laue photograph taken after the full conversion of the monoclinic selenium into the hexagonal form (c) are given in Fig. 4. More than twenty crystals were analyzed by Laue patterns and photomicrographs and it was concluded that, irrespective of the choice of the régime of heating, the monocrystal of monoclinic α- and β-red selenium always converts into polycrystalline hexagonal selenium.

As has been mentioned,[14] one of the basic causes of the conversion of a monocrystal to a polycrystal is the non-zero differences of densities of the matrix and the growing crystals ($\rho_{hexagonal} - \rho_{\alpha\text{-monoclinic}} = 0.44$ g/cm^3 and $\rho_{hexagonal} - \rho_{\beta\text{-monoclinic}} = 0.32$ g/cm^3).

Observations and photomicrographs of the α- and β-modifications of monoclinic selenium during conversion into the hexagonal form showed the growth of dendrites. The direction of the main trunk does not depend on the crystallographic direction of the matrix crystal. The branches of the dendrite make an angle of 58° with the main trunk and are strongly parallel between themselves. As seen in Fig. 2b the main trunks of the two dendrites are parallel and the branches of the second appear to be a continuation of the first branch. The dendritic growth of hexagonal selenium during the conversion of monocrystals of monoclinic selenium into the hexagonal form depends strongly on the temperature and rate of heating. In the given photomicrographs it is clearly seen that the space between the branches of the dendrite is filled by the generation of new nuclei in the line of branches. The monocrystal is eventually converted into a polycrystal.

The growth of dendrites in the case of the conversion of the α- and β-monocrystal of monoclinic selenium is the same as the dendritic growth of crystals from solutions, melts and by condensation of the vapor.

DISCUSSION

It is known that molecules of α- and β-monoclinic selenium are eight-atom closed rings. In the unit cell of the two phases there are four molecules. The monoclinic α-modification of selenium is different from the β-form in the packing of molecules, as is seen from the parameters of the unit cell (Fig. 5 a and b). In the ring of the molecule of α- and β-monoclinic selenium the distance of Se–Se is 2.34 Å and the shortest distance between the atoms of neighboring molecules of α-monoclinic selenium is 3.53 Å (the distance for the β-form being 3.48 Å).

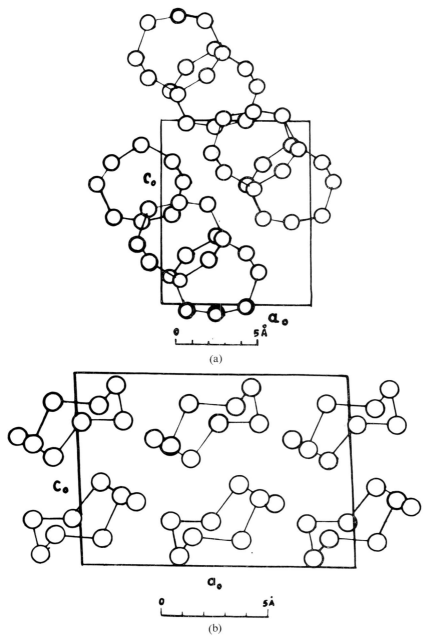

FIG. 5. (a) The projection of the monoclinic structure of α-selenium along the axis b_0; (b) The projection of the monoclinic structure of β-selenium along the axis b_0.

A molecule of hexagonal selenium consists of atomic chains displaced in zig-zag fashion in the direction [0001] (Fig. 6). The shortest interatomic distance in the chain and the shortest distance between the chains of hexagonal selenium are 2.32 Å and 3.53 Å respectively. The covalent bond angle Se–Se–Se of α- and β-monoclinic and hexagonal selenium is 105°. Between the atoms in the rings and chains in the case of α- and β-monoclinic and hexagonal selenium, there exist homopolar links and the rings and chains themselves are bound one to another by considerably weak non-polar forces.

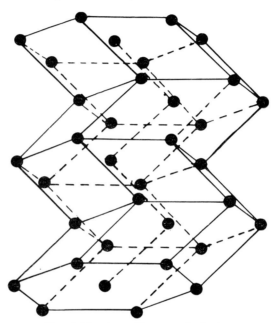

FIG. 6. Structure of hexagonal selenium.

As distinguished from the previous investigations,[14] the conversions of the α- and β-monoclinic monocrystal of selenium into the hexagonal form takes place in two stages. In the conversion of $Se_{monoclinic} \rightarrow Se_{hexagonal}$ a primary stage in the process may be the breaking of one of the Se–Se bonds of the eight-atom rings with the building up of the chains. Then an opening of the ring and generation of the eight-atom chain molecule with free radical-like ends takes place.

When the n-rings are broken, the n Se–Se bonds are destroyed. These short chains unite and form an infinite chain and reform $(n-1)$ Se–Se bonds. As n is very great, it can be considered equal to $(n-1)$ and it may be said that the number of the broken and newly originated bonds is equal.

After such an initial stage the nuclei of hexagonal selenium originate. As a rule, nuclei of the new phase are always originating at an imperfection (dislocations, vacancy, etc.) where the molecules of the matrix crystal are misplaced.

During heating of the monoclinic monocrystal of selenium the amplitude of thermal vibrations of atoms in the molecule and the amplitude of thermal vibrations of the molecule itself are increased.

At the imperfect section of the crystal the molecule has more freedom and energy, so there the molecule deforms. In the deformed closed eight-membered molecule of the matrix crystal, in a certain place, the Se–Se bonds are broken. These distorted molecules, probably, are the nuclei of hexagonal selenium. Since the intermolecular distances (i.e. intermolecular bonds) along the different crystallographic directions are different, the work necessary for the displacement of the molecule along these directions is also different. Therefore, it may be expected that the growth of crystal in the solid monocrystalline state should be peculiar to the anisotropy, which is characteristic of the physical properties of the crystal. The anisotropy must be more pronounced, if the symmetry of the crystalline lattice of the matrix crystal is low. If not, the anisotropy of the matrix monocrystal would always direct the nuclei of the new phase in some crystallographic direction, i.e. in the case of selenium it would take place as a strong bond between the growing and initial phases. Since the nuclei of hexagonal selenium originate from the monocrystal of monoclinic selenium in arbitrary directions and grow as dendrites, the strong bond between the growing and initial phases is absent. Also the investigation of the polymorphism in the low-symmetrical monocrystal,[14] where a monocrystal converts into a monocrystal, showed that the anisotropy of the matrix monocrystal has no influence on the internal growth in the crystal of the new phase. Thus the initial phase provides the new growing crystal with an isotropic medium. This phenomenon is, probably, connected with the fact that the heat of crystallization is released during the initial nucleation and growth of the crystal of the new phase inside the non-equilibrium phase. The heat released during nucleation and growth can go out into the surrounding medium, i.e. into the volume of the matrix crystal. In the case of selenium the greater part of the heat of crystallization is used to break the Se–Se bond in the molecule Se_8, and a part of it will assist in the annihilation of the spatial anisotropy of the matrix crystal but here the anisotropy of the growth velocities of the growing crystal is observed.

Since the atom chains in the hexagonal modification of selenium are placed spirally in the direction of [0001] and inside the chains between the atoms, the homopolar bonds act and the chains are interconnected by correspondingly weak forces, then the velocity of growth in the direction of [0001] must be especially great. More intensive growth of the nuclei of hexagonal selenium in the direction of [0001] promotes heat assumulation (selenium possesses low heat conductivity) in the bulk of the matrix crystal and creates conditions for the origin of dendrites.

Since $\rho_{hexagonal} > \rho_{monoclinic}$ hexagonal selenium crystal growth inside the monoclinic monocrystal of selenium is terminated because of the contact disturbance between the growing crystal and the crystal medium. Therefore, the conversion of $Se_{monoclinic} \rightarrow Se_{hexagonal}$ is continued only at the expense of the origin of new centres of crystallization at the phase boundary and, as a result, the monoclinic monocrystal of selenium is converted into a polycrystal of hexagonal selenium. It is also seen from the photomicrograph in Fig. 2c that the basic trunks and the initial branches of the dendrite of hexagonal medium are composed of a number of different sized blocks. This is also apparent from the Laue pattern in Fig. 4c which was obtained from the sample after the full conversion of the monoclinic monocrystal of selenium into the hexagonal form.

From the Laue pattern in Figs. 6 a and b, obtained under similar conditions ($T = 20\,°C$) it is seen that during the growth of the crystal of hexagonal selenium inside the monoclinic monocrystal of selenium the matrix crystal deforms slightly.

Both the α- and β-forms of monoclinic selenium differ from one another only by the orientation and the packing of the molecules. Symmetry and crystallographic unit cells are similar. Both modifications, therefore, must have almost equal energies and it is assumed that their mechanism of conversion and the growth form of the new phase will be similar.

During the conversion of $Se_{monoclinic} \rightarrow Se_{hexagonal}$, the low heat conductivity of the matrix monocrystal, the great amount of latent heat, the rapid growth in the direction of [0001] and the ease of nucleation of the hexagonal selenium inside the monoclinic monocrystal of selenium lead to dendritic growth.

THE GROWTH KINETICS OF THE HEXAGONAL SELENIUM CRYSTAL FROM THE MONOCLINIC SELENIUM MONO- CRYSTAL DURING STRUCTURAL CONVERSION

The kinetics and mechanism of the conversion of $Se_{amorphous} \rightarrow Se_{hexagonal}$ have been investigated by many authors.[1-3] In spite of the abundance of experimental investigations these questions have not been solved satisfactorily. Until the present study, the kinetics and mechanism of the conversion of $Se_{monoclinic} \rightarrow Se_{hexagonal}$ had not been investigated.

In the amorphous selenium about equal amounts of Se_8 and Se_x are present during crystallization. Depending on the conditions, different combinations of rings and spirals may be obtained. Therefore, depending on the conditions of the experiments and the history of the sample, the values of the physical properties of grey selenium are not constant. It is likely that at temperatures above 180°C the crystallized polymeric grey selenium has a homogeneous chain-like structure. Proceeding from the molecular point of view amorphous selenium presents a product of an intermediate stage of polymerization. It is supposed that the glassy black selenium is nearer to crystalline grey selenium, and the amorphous red selenium is nearer to the crystalline red selenium. The red crystalline form of selenium originates only from the molecule of Se_8 with the monomeric structure. The conversion of a red selenium monocrystal into grey selenium is the only case in which the crystallization to the hexagonal modification of selenium takes place at the expense of the breaking of rings of the eight-atom molecule of monoclinic selenium and of the production of the long chains.

EXPERIMENTAL

All the monocrystals chosen for this study had approximately similar dimensions of $1 \times 1 \times 0.3$ mm³.

Since the crystals of hexagonal selenium grow inside the monocrystals of α- and β-monoclinic selenium in the shape of dendrites (Fig. 2c) and there is no flat interface between phases during the growth of hexagonal selenium (e.g. conversion is a monocrystal-polycrystal) we cannot utilize the method described previously[15] where the velocity of the face (hkl) of the growing monocrystal inside the matrix monocrystal was measured. Therefore, the velocity of transition to the new polymorphic modification was measured with the crystal surface in the viewing field of a microscope.

All the monocrystals of monoclinic selenium chosen for this study had the form of thin plates, the dimensions of which were less than the area of the field of view. Thus it was possible to measure the exact area of each monocrystal. It is necessary to note that the thickness of the crystals used were almost similar (~ 0.3 mm). At each temperature the time from nucleation until the full conversion of the matrix monocrystal was measured. Thus, at each constant temperature the velocities of conversion of four crystals were measured.

The experimental data obtained were subjected to further treatment. The curves of $V_{monoclinic} \to V_{hexagonal}$ vs. temperature (Fig. 7a) showed a linear dependence of $\ln V$ vs. $1/T$ (Fig. 7b).

DISCUSSION

Published reports about the boundaries of stability of α- and β-monoclinic modifications of selenium are contradictory. For instance, Costy[16] gives an upper temperature limit for the existence of the monoclinic modification of

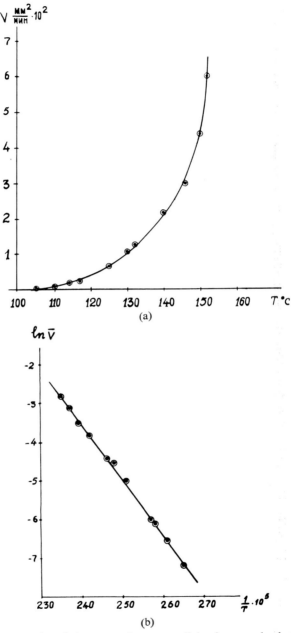

FIG. 7. (a) Velocity of the conversion, monoclinic→hexagonal selenium, as a function of temperature. (b) Logarithm of the velocity of the conversion, monoclinic→hexagonal selenium, as a function of $1/T$.

105–107.5°C. Mondain-Monval[17] considered that this modification is stable below 120°C. Muthmann[18] stated that the first monoclinic modification is transformed into the hexagonal one at 110–120°C, and the second one at 125–130°C. The apparent temperature of the equilibrium conversion of α- and β-monoclinic selenium into the hexagonal form depends only on the heating rate and the rate of nucleation and growth in the temperature interval between 115 to 130°C. Further investigation showed that the conversion of $Se_{monoclinic} \rightarrow Se_{hexagonal}$ has no specific phase equilibrium temperature. The nuclei of hexagonal selenium inside the monoclinic monocrystal of selenium and their growth depend on the temperature and the time. For instance at 90°C, conversion of $Se_{monoclinic} \rightarrow Se_{hexagonal}$ continues for weeks. At 95°C, $V \approx 10^{-6}$ mm²/min and at 105°C, the velocity of conversion reaches $V = 0.7 \times 10^{-4}$ mm²/min.

It seems that rate of conversion is connected with the instability of the molecule of monoclinic selenium. The ring molecule of Se_8 of monoclinic selenium is easily converted into the chain molecule of Se_x of hexagonal selenium ($nSe_8 \rightarrow mSe_x$).

The linear velocity of growth of hexagonal selenium crystals inside the monoclinic selenium monocrystal may be given by the empirical expression,

$$V = A \exp^{-E/RT}.$$

Here $A \sim \exp(\Delta S/R)$, when ΔS is the difference of the entropies of the phase and of the transition state. If we assume that pressure plays no part in the conversion of monoclinic selenium into the hexagonal form, the physical scheme for the conversion is as follows:

The phase and the transition state energies are not changing in the temperature interval under consideration. It may be assumed that the difference of entropies of phases and of the transition state are weakly dependent on the temperature.

From the direct dependence of ln V on $1/T$ the activation energy value $E = 27,810$ cal/mole and $A \approx 10^{13}$ mm²/min were obtained. These values are very high. The activation energy, for example, is definitely well above that for self-diffusion and much higher than for grain boundary diffusion. It may be explained as follows:

The growth of the hexagonal selenium crystal inside the monoclinic monocrystal of selenium is a phase conversion of the first type. It is known that such conversions are accompanied by a change of density. Because of the great difference in density between hexagonal selenium and monoclinic selenium, growth of the hexagonal form inside monoclinic selenium causes the appearance of internal stresses. Therefore, it may be supposed that between the boundary of the growing crystals and the crystal-medium there is a deformed (compressed or extended) layer of molecules. In these deformed eight-atom closed rings of the monoclinic monocrystal of selenium, as mentioned above, one of the Se–Se bonds is terminated and the newly formed chain molecules, with the radical-like ends, are the medium providing the further growth of the crystal of hexagonal selenium.

PHYSICAL PROPERTIES OF THE MONOCLINIC MONOCRYSTAL OF SELENIUM

Selenium is widely used for the manufacture of photocells, rectifiers, electrophotographic plates, etc. In these devices hexagonal and amorphous selenium are used.

It is known that in addition to the amorphous and hexagonal modifications there exists a monoclinic modification of selenium. For some reason investigators up to now have not paid due attention to the monoclinic modification. Comprehensive investigations of the physical properties and phase stability would define the possibility of its application in the semiconductor industry.

Kyropoulos[7] measured with red light, $\lambda = 0.620\ \mu$, the optical constants, i.e. the index of refraction $n = 2.2–2.9$ and the dielectric constant $\epsilon = 7.39$. Prosser[4] measured optical constants near the absorption edge at $\lambda = 0.876–0.592\ \mu$, $n = 2.47–2.74$. Also Prosser[5] and Cudden and Pohl[6] measured the photoconductivity spectral distribution of the α-monoclinic selenium monocrystal.

The present work is the first in a series of investigations devoted to the elucidation of the physical properties of the monoclinic selenium monocrystals.

Determinations of the electrical properties of the monoclinic selenium monocrystal were carried out on samples (Fig. 1) of $3 \times 3 \times 2$ mm^3 dimensions. Vacuum deposited bismuth and nickel contacts were used in this study.

The monoclinic selenium monocrystals at room temperature had a specific resistance of the order of 10^9 ohm-cm.

The calculated value of the carrier concentration was 10^{12} cm^{-3} (the mobility is taken as 10^{-1} cm^2/volt-sec).

The monoclinic selenium monocrystals were found to possess considerable photoconductivity in the visible region of the spectrum. The ratio of the dark to light resistance at a field of 500 volt/cm was $10–10^2$.

Figure 8 (curve 1) shows the spectral distribution of the photoconductivity of the monoclinic selenium monocrystal. It is seen that the maximum of photosensitivity corresponding to the self-absorption of the monoclinic selenium lies in the violet region, $\lambda_m = 0.450\ \mu$. The band gap, calculated by the half maximum point of photoconductivity, is 2.4 eV and agrees with the results determined by the absorption edge.[4]

According to the literature data,[4, 5] the absorption edge of monoclinic and of hexagonal selenium is different. The absorption edge of hexagonal selenium is displaced to the long wavelength side (Fig. 9).

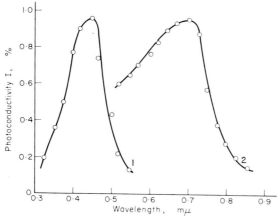

FIG. 8. Spectral distribution of photoconductivity : 1. Monocrystal of monoclinic selenium ; 2. Polycrystal of hexagonal selenium, after conversion of $Se_{monoclinic} \rightarrow Se_{hexagonal}$.

A determination of the spectral distribution of photoconductivity shows the same displacement, with the change of the selenium modification, as absorption curves.

Figure 8 (curve 2) shows the spectral distribution of photoconductivity of the same sample taken after heat treatment at 140°C for 2 hr. It is seen that after heat treatment, a result of which is the conversion of the monoclinic modification of selenium into the hexagonal form, the curve of the spectral distribution of photoconductivity is displaced to longer wavelengths. This

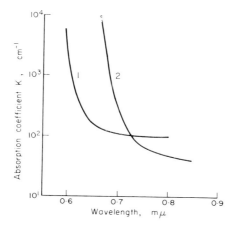

FIG. 9. Absorption curves. 1. Monocrystal of monoclinic selenium. 2. Monocrystal of hexagonal selenium.

displacement is about 0.25 eV. The maximum of photoconductivity ($\lambda_m = 0.7\,\mu$), the band gap calculated by the half maximum value of photoconductivity (1.7 eV), the electrical conductivity (10^{-4} ohm^{-1} cm^{-1}) and the density of carriers (10^{17} cm^{-3}) agree with the corresponding data for polycrystalline hexagonal selenium.

Figure 10 shows the spectral distribution of photoconductivity for the sample which was treated thermally at 140°C for 1 h. As can be seen, there are the two maxima which relate to the monoclinic and hexagonal modifications of selenium. From Fig. 10 it can also be seen that the sample, at the

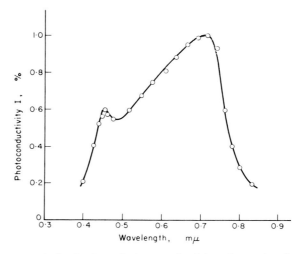

FIG. 10. Spectral distribution of photoconductivity of sample of monoclinic selenium converted partially to the hexagonal form.

mentioned régime of thermal treatment, has not been transformed completely into the hexagonal modification. There exists a definite percentage of the monoclinic modification in the hexagonal selenium.

On the basis of the above data it is possible to suggest that in the experiments by different authors[5, 6] the two maxima observed in the spectral characteristics of selenium are specified by the presence of hexagonal material in the monoclinic component.

The above-mentioned results can be interpreted by the supposition that the peak of photoconductivity at $\lambda = 0.45\,\mu$ corresponds to the transition of carriers inside the rings, and at $\lambda = 0.7\,\mu$ it corresponds to the transition between the chains. According to the theoretical model[19] the distribution of atoms in the monoclinic selenium creates rings and the transition of carriers takes place mainly inside the rings. In hexagonal selenium the molecule consists of chains and the transition of carriers takes place along them.

In the case of monoclinic selenium there is no maximum of photoconductivity which corresponds to the transition between the chains.

REFERENCES

1. Kozyrev, P. T., *Zh. Tekhn. Fiz.* **28**, 500 (1958).
2. Bolotov, I. E. and Muravyev, E. A., *Fiz. Tverd. Tela.* **8**, 1585 (1966).
3. Blet, G., *J. Phys. et Radium* **22** (2), 17 (1961).
4. Prosser, V., *Cesk. Casopis Fys.* **10**, 35 (1960).
5. Prosser, V., *Proc. Int. Conf. Semiconductor Physics*, Prague, 1960, Academic Press, New York, 1961.
6. Cudden, B. and Pohl, R., *Z. Physik*, **35**, 243 (1925).
7. Kyropoulos, S., *Z. Physik*, **13**, 618 (1926).
8. Bradley, A. J., *Phil. Mag.* **48**, 477 (1924).
9. Burbank, R. D., *Acta Cryst.* **4**, 140 (1951).
10. Burbank, R. D., *Acta Cryst.* **5**, 236 (1952).
11. Marsh, R. E., Pauling, L. and McCullough, J. D., *Acta Cryst.* **6**, 71 (1953).
12. Chizhikov, D. M. and Schastlivyi, V. P., *Selenium and Selenides*, Nauka, Moscow, 1964; English translation, E. M. Elkin, Collet's Limited, Wellingborough, England, 1968.
13. Asadov, Yu. G., *Doklady Akad. Nauk SSSR*, **173**, 570 (1967).
14. Kitaigorodskii, A. I., Mnyukh, Yu. V. and Asadov, Yu. G., *J. Phys. Chem. Solids*, **26**, 463 (1965).
15. Mnyukh, Yu. V., Kitaigorodskii, A. I. and Asadov, Yu. G., *Zh. Eksperim. i Teor. Fiz.* **48**, 19 (1965).
16. Costy, M., *Compt. Rend.* **149**, 674 (1909).
17. Mondain-Monval, P., *Bull. Soc. Chim.* s4, **39**, 1349 (1926).
18. Mellor, J. W., *A Comprehensive Treatise on Inorganic Theoretical Chemistry*, Vol. X, Longmans, London, 1940, p. 706.
19. Caspar, R., *Festkor. Phys. and Phys. Leuchtstof*, Berlin, 1958.

PREPARATION AND IDENTIFICATION OF SELENIUM α-MONOCLINIC CRYSTALS GROWN FROM SELENIUM-SATURATED CS$_2$

S. IIZIMA, J. TAYNAI and M-A. NICOLET

California Institute of Technology, Pasadena, California

SOLID selenium is known to exist in four different modifications: amorphous, hexagonal (trigonal), α-monoclinic and β-monoclinic. When grown at room temperature from a solution of CS$_2$ saturated with selenium, the structure is monoclinic. According to the literature, both monoclinic modifications can be obtained by this method in the form of small single crystals of millimeter size at most.

Figure 1 shows the lattice vectors of α- and β-monoclinic selenium used by Mitcherlich[1] and Muthmann [2] to describe their goniometric observations. More recently Burbank[3] has proposed a different unit cell which relates to

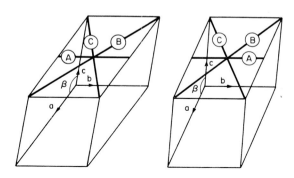

Ⓐ TRACE BETWEEN (001) & (100)   Ⓐ TRACE BETWEEN (001) & (100)
Ⓑ TRACE BETWEEN (001) & (111)   Ⓑ TRACE BETWEEN (001) & (110)
Ⓒ TRACE BETWEEN (001) & ($\bar{1}$11)   Ⓒ TRACE BETWEEN (001) & ($\bar{1}$10)

a : b : c = 1.63495 : 1 : 1.6095   a : b : c = 1.5916 : 1 : 1.1352
β = 104° 2'   β = 93° 4'

FIG. 1. Unit cells of α- and β-monoclinic crystals introduced by Mitcherlich[1] and Muthmann.[2]

the others as shown in Fig. 2. Correspondence between a few sets of Miller indices for the two systems are given in Table 1. For convenience the reference systems of Fig. 1 are used exclusively below.

TABLE 1. *Correspondence Between Miller Indices of Planes Described by the Unit Cells of Fig. 2*

Burbank's indices	Muthmann's indices
101	001
110	111
100	101
010	010
10$\bar{1}$	100
$\bar{1}$10	$\bar{1}$11
111	012
1$\bar{1}$1	0$\bar{1}$2
$a':b':c'=9.05:9.07:11.61$	$a:b:c:=1.63495:1:1.6095$
$\beta'=90°\ 46'$	$\beta=104°\ 2'$

According to Klug[4] it is almost impossible to distinguish α from β crystals by simple microscopic observation because the angle between the (100) and the (110) faces is 57° 46' for the α and 57° 59' for the β modification. On the other hand, Mitcherlich and Muthmann state that the principal faces of their

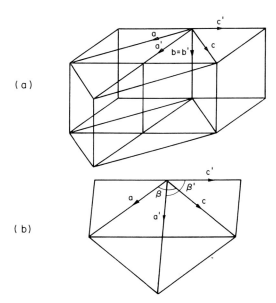

FIG. 2. Connection between the unit cell a', b', c' of Burbank[3] and that of Fig. 1.

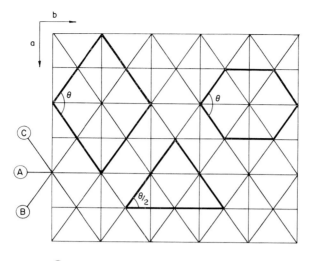

FIG. 3. Shapes of (001) faces, expected from the traces shown in Fig. 1.

monoclinic crystals are (001) planes and that these faces are usually intersected by facets identified as (100), (111) and ($\bar{1}$11) planes for α crystals and by (100), (110) and ($\bar{1}$10) planes for β crystals. Traces of these planes are shown on the upper (001) faces of Figs. 1 a and b. A top view of those traces extending over fifteen adjacent unit cells is reproduced in Fig. 3. Mitcherlich's and Muthmann's observations thus lead one to expect that the principal faces of monoclinic selenium crystals should preferentially develop hexagonal, rhomboidal or triangular boundaries. From Fig. 1, the value of θ is $2 \tan^{-1}(a/b)$, or 117° 6' for the α modification and 115° 42' for the β modification. The difference of 1° 24' is well within the power of resolution of a microscope equipped with a cross-hair ocular and a graduated rotating stage. It should therefore be possible to determine the crystallographic identity of monoclinic selenium single crystals if angles between boundary lines on a principal face rather than angles between faces are measured.

We have obtained selenium single crystals in the form of very thin platelets, which are approximately 1200 Å thick and hexagonal in shape. These crystals

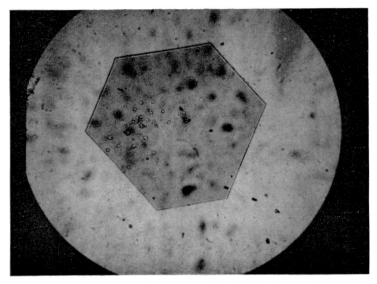

Fig. 4. Photomicrograph of an α-monoclinic selenium hexagonal platelet.

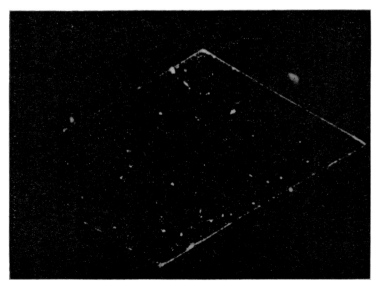

Fig. 5. Photomicrograph of an (001) face of a bulky α-monoclinic selenium crystal.

were grown by evaporating CS_2 saturated with selenium on a glass substrate at room temperature. They have a maximum diameter of several hundred microns. Figure 4 shows a photomicrograph of such a crystal. The average value of θ determined from seven such platelets is 116° 58' with an estimated error of 10', indicating that these platelets are α-monoclinic.

We have also grown bulk single crystals of millimeter size from a saturated solution of selenium in CS_2 with amorphous selenium of high purity as a source.[5] These crystals usually have at least one rhomboidal face or slight variations of it. Figure 5 shows an example. The angle θ was measured on five such faces. The average value is 117° 3'.

Finally, we have grown rhomboidal platelets on bulk crystals described above by evaporating CS_2 saturated with selenium on (001) faces of bulk crystals. Angles of two such platelets were measured and yielded $\theta = 117°\ 12'$. Triangular platelets were also observed in this case. However, their sizes were too small to permit accurate measurements.

It is concluded from the measured values of 116° 58', 117° 3' and 117° 12' for θ and the two possible values of 117° 6' for α- and 115° 42' for β-monoclinic selenium that the present crystals are of the α modification. The difference of 1° 24' is well within the resolving power of an optical microscope. The test based on angles between boundary lines of a face should thus provide a simple method to identify α- and β-monoclinic crystals of selenium. No β-monoclinic selenium crystals were found.

Two crystals so identified as α-monoclinic have been analyzed by X-rays. The results indicate a high degree of lattice perfection and agree with the lattice constants given by Burbank[3] for α-monoclinic selenium.

ACKNOWLEDGMENT

The high purity amorphous selenium was obtained through the courtesy of the Selenium–Tellurium Development Association, Inc. Dr. R. Marsh performed the X-ray study. We are thankful for this assistance.

REFERENCES

1. MITCHERLICH, M., *Ann. de Chimie et de Physique* **46**, 301 (1856).
2. MUTHMANN, W., *Z. Kristallogr.* **17**, 336 (1890).
3. BURBANK, R. D., *Acta. Cryst.* **4**, 140 (1951).
4. KLUG, H. P., *Z. Kristallogr.* **88**, 128 (1934).
5. IIZIMA, S. and NICOLET, MA., JPL Space Programs Summary 37–48, Vol. III, p. 70, (1967).

CRYSTALLIZATION AND VISCOSITY OF VITREOUS SELENIUM

H. P. D. LANYON

Department of Electrical Engineering, Worcester Polytechnic Institute, Worcester, Massachusetts 01609

INTRODUCTION

Selenium exists in two long-chain molecular forms in which the atoms are arranged in helical spirals, the repeat distance along the spiral being three atoms. The thermodynamically stable form at room temperature is the hexagonal form in which the spirals are arranged parallel to the crystallographic c-axis. An isotropic vitreous (or glassy) selenium can be produced either by quenching liquid selenium or by slow evaporation of selenium onto a substrate held at room temperature. In this form of selenium, although the short-range order is similar to that in hexagonal selenium, there is no long-range order: the whole may be thought of as a tangled mass of interpenetrating chains. If allowed to stand, the vitreous phase will convert to the hexagonal form, the rate of transformation depending markedly upon temperature. It has been shown both in our own work and in that of others[1,2] that the crystallization process takes place in two stages. Firstly a nucleus of crystalline selenium forms, followed by the growth of this crystallite. It is the initial nucleation that is the difficult process. Once nucleation has occurred, further crystallization normally takes place by the growth of the initial crystallites rather than by the formation of new ones. The density of such crystallites is independent of the temperature of crystallization and appears to be related to imperfections in the selenium layer. The rate of incorporation of new material into the crystallites is proportional to the surface area and is strongly temperature dependent. This temperature dependence of the crystallization rate has often been associated with the change of viscosity of vitreous selenium (supercooled liquid) with temperature. The viscosity has been measured close to room temperature by Mondain-Monval[3] and in the liquid form by Krebs[4,5] among other workers. Krebs also studied the effect of various impurities such as arsenic upon the viscosity. The viscosity was found to decrease from approximately 10^{14} poise at room temperature to about 20 poise at the melting point of selenium.

In this paper results of the measurement of viscosity close to room temperature and of the crystallizability of selenium alloys with tellurium and arsenic additions are presented. These results show that the simple criterion of the viscosity of the material cannot be used alone to predict the crystallizability of an alloy. There are a number of differences between the results reported here close to room temperature and those obtained by Krebs on similar materials at higher temperatures.

EXPERIMENTAL

(a) *Crystallization*

The crystallization rates of the selenium and its alloys were measured using a differential thermal analysis technique.[6] Approximately 1 g of the material was sealed in an evacuated quartz ampoule, heated above the melting point and then quenched rapidly to room temperature or below. It was then heated at a constant rate of 3 °C/min in a furnace and the rate of increase of sample temperature compared with that of a dummy aluminum oxide sample. The differential temperature was then recorded directly as a function of the

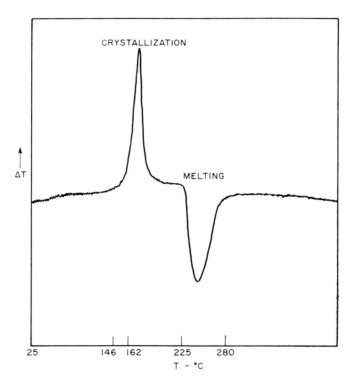

FIG. 1. Typical DTA trace for the crystallization of selenium.

furnace temperature. An exothermic process such as crystallization shows up as an increase in the temperature of the sample. Conversely, with an endothermic process such as melting, the sample temperature lags behind that of the dummy. A typical DTA trace is shown in Fig. 1 for a pure selenium sample. If arsenic is added to the selenium, the temperature at which the peak appears increases with increasing arsenic content. On the other hand, with tellurium the crystallization temperature falls with the increased tellurium content.

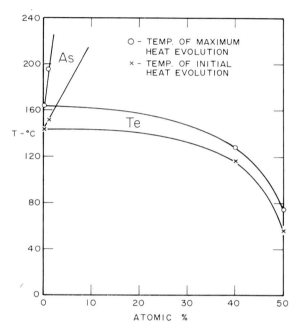

FIG. 2. Concentration dependence of crystallization temperature of selenium/tellurium and selenium/arsenic alloys.

Figure 2 shows the concentration dependence of the crystallization temperature as measured by the DTA technique at constant heating rate. The temperature at which the evolution of heat first appears has been taken as a measure of the crystallizability of the alloys studied in this work. This temperature is a function of the heating rate and should not be taken as an absolute temperature at which crystallization first occurs. For example, in the case of pure selenium, crystallites form slowly at temperatures as low as 50°C. The crystallization temperature found by the DTA technique is 143°C. However, since all of the measurements have been taken at a constant heating rate of 3°C/min, there will be a correlation between the ease of crystal formation and the temperature estimated from the DTA measurements.

(b) *Viscosity*

The viscosity measurements were made using the point-indentation technique originated by Mondain-Monval.[3] Samples were prepared by heating the selenium or alloy in air above its melting point in a 1-in. diameter hole punched in $\frac{1}{8}$-in. aluminum sheet. Similar aluminum sheets were placed on either side and the whole quenched in cold water. The two retaining sheets were then removed and a 0·001-in. stainless-steel point loaded with 100 g

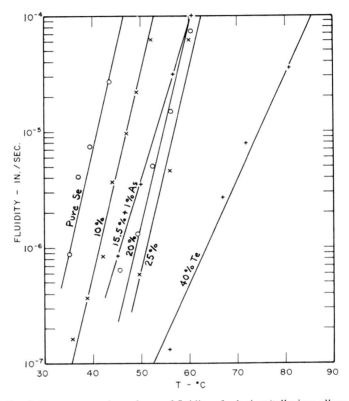

FIG. 3. Temperature dependence of fluidity of selenium/tellurium alloys.

placed on the free sample surface. The creep was measured as a function of time using a strain gauge. The whole apparatus, including the strain gauge, was immersed in an oil thermostat in order to avoid thermal gradients. Once the apparatus had reached thermal equilibrium, the point was lowered onto the surface and the creep recorded. The results for pure selenium agree very closely with those of Mondain-Monval over the same temperature range. No attempt was made to calibrate the apparatus absolutely in order to check the accuracy of Mondain-Monval's results. A fitting to the known value of viscosity of selenium at room temperature would normalize all of the curves shown. Measurements were always repeated at a lower temperature after the temperature had been taken to its maximum to check that no irreversible change in viscosity had taken place during the measurement cycle, invalidating the results.

Figures 3 and 4 show the effect on the fluidity ($\alpha 1/\text{viscosity}$) of increasing amounts of tellurium and arsenic respectively. In Fig. 3 it can be seen that the effect of the tellurium is to increase the viscosity of the alloy without changing the general shape of the curve. At any particular temperature the change in

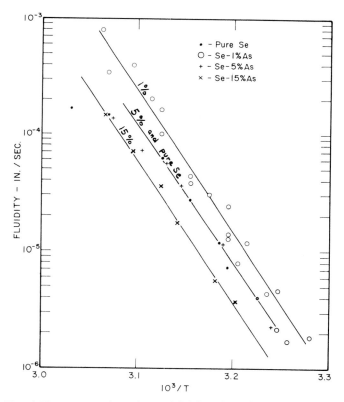

FIG. 4. Temperature dependence of fluidity of selenium/arsenic alloys.

viscosity appears to be directly proportional to the tellurium content within the experimental accuracy. It is of interest, in view of the results that will be presented for arsenic, to note that one of the curves is for a sample containing selenium, tellurium and arsenic in the atomic proportions 84 : 15 : 1. This curve lies in essentially the place that one would expect the curve for an alloy without arsenic.

Figure 4 shows the results obtained for alloys with arsenic included. Initially the effect of arsenic appears to be to *reduce* the viscosity. With 5 atomic per cent arsenic addition the viscosity is the same as for pure selenium. With higher percentages the viscosity does increase, but slowly. The increase is by no means comparable with that observed with the tellurium addition.

DISCUSSION

In the present experiments it has been found that arsenic and tellurium affect the crystallization rate of selenium in the same way as has been observed previously by other workers. The effects of the two elements appear to be additive. The temperature of crystallization measured was typically in the range 56–180°C. Previous measurements of viscosity in the selenium–arsenic system were taken above the melting point. In this work the viscosity was measured in the temperature range between 30° and 80°C, which is the temperature range of interest in the slow growth of crystallites in the selenium–tellurium alloy system. Previous measurements have shown the growth of crystallites in this temperature range and the slow change in electrical resistance[7] as the crystallites grow. The rate of conversion is exponentially dependent on the temperature.

One of the reasons that has been suggested for the temperature dependence of the crystallization rate has been that at low temperatures the viscosity of the selenium is too high for appreciable bulk transfer of the selenium to the crystallite/liquid interface. The effect of arsenic on the crystallization rate was attributed to cross-linking between neighboring chains which caused an increase in viscosity. Conversely, tellurium and halogens increased the crystallization rate by shortening the average chain length, thus decreasing the viscosity. The measurements of Krebs on the viscosity of selenium with halogen or arsenic additions were consistent with this hypothesis.

The present measurements suggest that this simple picture of crystallization is not applicable. In the temperature range in which the growth of small crystallites occurs, the addition of arsenic does not alter the viscosity of selenium appreciably, although it has a marked effect on the crystallization rate. Conversely, the effect of tellurium is to increase the viscosity and not to decrease it as the simple theory would predict. Similar measurements of the effect of iodine additions again showed no large change in viscosity up to a 1 per cent addition. At higher concentrations there appeared to be some decrease in the viscosity. The latter results are complicated by the tendency of the alloy to crystallize. The DTA measurements on the iodine alloys did not show a decrease in the temperature at which the peak emission of heat occurred. However, there was a marked increase in the width of the peak, the initial rise in temperature occurring at a lower temperature.

The addition of dopants to the selenium not only changes its viscosity but also affects the ease with which the individual chains can attach themselves to the crystal surface to contribute to the growth of the crystal. In the present case, the main effect of the incorporation of arsenic into the chains is not to cross-link (this can only occur when more than one arsenic atom is associated) but to cause chain branching. Since this causes disorder in the crystal it is geometrically and hence energetically less favorable for crystallization to occur. Conversely, with thallium and the halogens or alkalis the chain length is shortened, easing placement of the chain segment. In the case of tellurium dopant the intermolecular forces aligning the chains are much stronger than for selenium and more than counteract the effect of increased viscosity so that the rate of crystallization increases with increasing tellurium content. The mechanisms outlined here apply equally to the growth of spherulites and single crystals. The main effect of a shorter chain length or greater intermolecular forces would be to enhance the growth of single crystals compared to spherulite formation since the probability of incorrect atomic placement is diminished. This is in agreement with observation.

The temperature dependence of the viscosity of all the alloys was very similar except for the value of the scaling parameter. This can be seen in Fig. 4 which is a plot of log (fluidity) vs. $1/T$ for the arsenic alloys. The plots are parallel straight lines suggesting that the mechanism of change in viscosity is the same in all of the alloys. The behavior for tellurium and iodine additions is similar. The activation energy derived from the slope of the plots is 2·6 eVs.

ACKNOWLEDGMENT

The work described in this paper was performed whilst the author was a Member of Technical Staff at the RCA Laboratories David Sarnoff Research Center, Princeton, New Jersey. The author would like to thank Dr. E. F. Hockings for assistance with some of the DTA measurements.

REFERENCES

1. KECK, P. H., *J. Opt. Soc. Am.* **41**, 53 (1951).
2. HASHIMOTO, K. *et al.*, *Mem. Fac. Sci. Kyusu Univ.* B **1**, 151 (1955).
3. MONDAIN-MONVAL, P., *Ann. Chem.* **3**, 18 (1935).
4. KREBS, H. and MORSCH, W., *Z. Anorg. Chemie* **263**, 305 (1950).
5. KREBS, H., *Z. Anorg. Chemie* **265**, 156 (1951).
6. LANYON, H. P. D. and HOCKINGS, E. F., *Phys. Status Solidi* **17**, K185 (1966).
7. LANYON, H. P. D., *J. Appl. Phys.* **35**, 1516 (1964).

GROWTH, PERFECTION AND DAMAGEABILITY OF TELLURIUM SINGLE CRYSTALS

E. D. Kolb and R. A. Laudise

Bell Telephone Laboratories, Incorporated, Murray Hill, New Jersey

INTRODUCTION

The discovery of the 10.6 micron CO_2 laser[1] makes it important to perfect modulator and harmonic generator materials with transmission in this region. Most conventional non-linear optical materials are oxides with charge transfer bands which make them opaque in the region beyond 10 microns. Among the more promising materials with appropriate transmission is tellurium, which is essentially transparent from about 5 microns to beyond 25 microns.[2] Tellurium also exhibits a large positive birefringence[2] which makes phase matching easy. Patel[3,4] has used tellurium in a harmonic generator and a tunable parametric amplifier, and has shown that its non-linear optical coefficients are large. For laser experiments at 10 microns, high transmission, low scattering and homogeneity in the index of refraction are required. Parametric oscillator experiments especially require crystals of high quality,[4] and this paper reports how such crystals are prepared.

Commercially available tellurium crystals were found to be too opaque in the 10 to 20 micron region for successful parametric experiments. Tellurium crystals have been prepared by the Stockbarger–Bridgman method,[5,6] Czochralski growth[7-9] and vapor growth.[10] In view of the ease with which large size, low dislocation crystals of semiconductors such as silicon and germanium have been obtained with the Czochralski method, it was decided to use that technique.

EXPERIMENTAL

The crystal puller and heater arrangement are shown in Fig. 1A. The temperature control system consisted of a nichrome resistance wound furnace fed by a saturable core reactor controlled from a 0–1 mV Leeds and Northrup CAT recorder controller. A 0–25 mV Zener diode bias supply was used to suppress the thermocouple voltage to the center of 0–1 mV scale. A constant

Fig. 1A. Crystal puller and heater arrangement for tellurium growth.

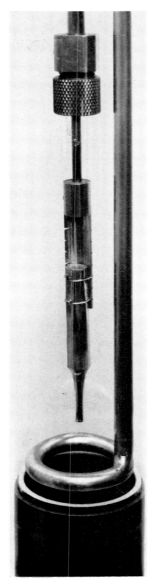

FIG. 1B. Gas distribution ring.

resolution potentiometer of the 0–25 mV bias supply was driven by a clock motor, through an interrupter timer, for automatic temperature programming. The system was capable of ±0.25°C control at 450°C (m.p. of Te). The crucible material was fused quartz in a graphite susceptor. The arrangement for vacuum sieving[11] is shown in Fig. 2. The procedure involved placing a ~140 g charge of "as supplied" Cominco† 6–9's tellurium in the upper crucible. The furnace tube was secured to the crystal growing machine, and it was evacuated

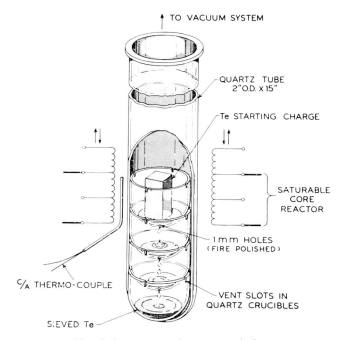

FIG. 2. Arrangement for vacuum sieving.

with a mechanical pump to the 3–10 micron range. By properly positioning the two bottom furnace windings the tellurium charge was melted and flowed to the next crucible. With additional furnace repositioning the charge was sieved through the three crucibles being finally collected in the fourth. This served to remove almost completely the TeO_2 which is found in most bulk tellurium. Approximately 120 of the original 140 g charge was recovered. The balance had resublimed above the crucibles and probably contained any impurities vaporized during the sieving and was therefore discarded. Vacuum sieving was found to be essential in removing TeO_2 before growth. Seeds were obtained originally by nucleating on selected polycrystalline pieces broken from

† Cominco Products, Inc., 818 West Riverside Ave., Spokane, Washington.

commercial tellurium ingots. Careful necking down to about $\frac{1}{16}$ in. diameter cross-section of new seed growth will result in propagating a single crystal which can then be enlarged again to diameters of ~ 0.600 in., in the present apparatus. Single crystals ~ 0.250 in. in diameter up to 3 in. long, obtained from Henley and Co., Inc.,† were also used as seeds. These contain enough imperfections, such as lineage and twins, that necking down of new seed growth is required to eliminate such defects in the attempt to grow enlarged single crystals. A typical growth run for optimum quality was carried out as follows: the hydrogen gas was tank H_2 passed through a Deoxo purifier,‡ an activated alumina trap (Lectrodryer)§ and a stainless-steel trap containing Linde molecular sieves (5A)†† in liquid N_2. The least amount of TeO_2 was observed when H_2 dew points were below $-80°C$. The growth chamber was maintained at 10 psi with a flow rate of 750 cm^3/min. A stainless-steel gas distribution ring (Fig. 1B) was used to direct the gas to the melt since a closed end furnace tube was used. The tube also served as a nucleation site for evaporating tellurium. Prior to its use the furnace tube became so coated from the deposited tellurium that the view of the growing crystal was badly restricted.

The seed orientation was always $\langle 0001 \rangle$, and rotation rates of 6 and 10 rpm were used. The slower speed is more desirable for the large diameter crystals where the cross-section is hexagonal. Eddy currents are observed on the surface of the melt even at 6 rpm. The eddy currents produce standing waves on the surface of the melt. Thus, the molten tellurium at the liquid–solid interface is oscillating up and down resulting in some uncontrolled growth variations. The pull rate was varied between 0.1–1.0 in./hr. It was found that the crystals grown at the faster rates contained more lineage and were of poorer quality than those grown at the slower rates. The conditions used were essentially those described by Keezer[9] except that the furnace tubes used were not tin oxide coated so that they could be resistance heated. By heating the furnace tube tellurium sublimation could be prevented. It is possible to see minute sublimed tellurium particles following the H_2 gas flow pattern in the furnace tube. The adherence of these on the growing crystal are very definite sites for defect formation.

CHARACTERIZATION AND DISCUSSION

Figure 3 shows two crystals, typical in size of those produced. The crystal on the left (Fig. 3A) exhibited lineage. Scattering at grain boundaries in such crystals made them useless in parametric experiments. Gross lineage of this

† Henley and Company, Inc., 202 East 44th Street, New York, N.Y.
‡ Baker Platinum, Division of Engelhard Industries, Newark, New Jersey.
§ Pittsburgh Lectrodryer Corp., P.O. Box 1766, Pittsburgh, Pennsylvania.
†† Union Carbide Corp., Linde Division, Rte 38, Morristown, New Jersey.

 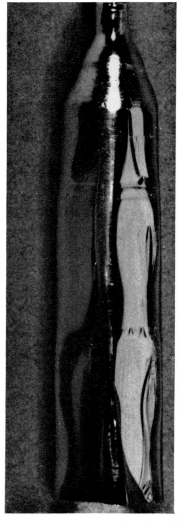

Fig. 3a. Tellurium crystal exhibiting lineage.   Fig. 3b. Tellurium crystal of good quality.

sort could be minimized by keeping the TeO_2 contamination to a minimum and by using slow seed rotation and pull rates. The crystal on the right (Fig. 3b) is typical of the good crystals produced. The ripples on the surface are due to surface lineage and did not persist through the crystal. The crystals of Fig. 3 (a) and (b) are about 3 in. long.

Early in this work it was noted that severe damage occurred in tellurium crystals that had been cut by spark erosion† and had been lapped with AO 305 and polished with Linde A. Figure 4B is a Laue picture of a crystal of the quality of Fig. 3B indicating poor crystallinity after the above treatment. Etching damaged crystals in 1 part H_2O : 1 part H_2SO_4 (freshly mixed) for

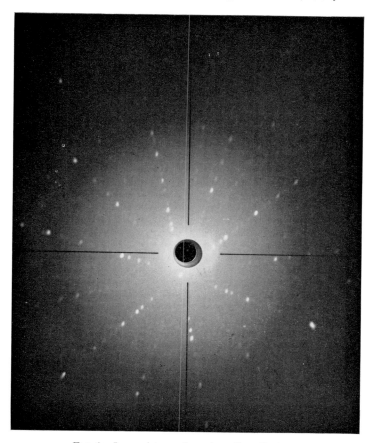

FIG. 4A. Laue picture of good-quality tellurium.

90 min showed most of the damaged layer was removed as evidenced by Laue picture Fig. 4A. Chemical string sawing, using a multistrand endless Saran string and 25 ml. saturated CrO_3–25 ml. concd. HCl–75 ml. H_2O, resulted in no damage detectable in Laue pictures.

Carrier concentrations were measured by the thermoelectric effect and were as low as $10^{15}/cm^3$ in the best crystals at room temperature. Crystals with

† Crystals were cut with Servomet Spark Machine, Metals Research Ltd., Cambridge, England.

carrier concentration much above $10^{16}/cm^3$ were not useful for parametric experiments. It is probable that reduced carrier concentrations were the result of vacuum sieving (which volatilized impurities) that was used for crystal preparation in the later stages of this work. Vacuum distillation of starting material tellurium would be expected to cause further reduction in carrier concentrations.

FIG. 4B. Laue picture of tellurium exhibiting lineage.

Infrared absorption measurements were made on a Cary Model 421. Figure 5 is a plot of the absorption coefficient α (corrected for reflectivity) versus wavelength for two tellurium crystals. The data for each of the curves were obtained on two specimens of different thickness and as can be seen, the agreement is good. Te 38K was not satisfactory for parametric amplifier measurement and Te 44K, which was satisfactory, is the best crystal grown in this equipment at this time. Parameters which both improved physical perfection and which reduced carrier concentration were found to improve infrared transmission.

Finally, crystals with carrier concentration below 10^{15}, whose morphology was like Fig. 3B, whose Laue picture was like Fig. 4A, and which were prepared by string sawing and/or etching away damage, were used in parametric oscillator experiments. Such crystals gave a gain of 3DB at 17.888 microns and crystals of lower quality were not useful.

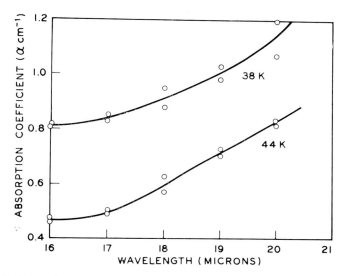

FIG. 5. Absorption spectrum of tellurium. 38K—not satisfactory for parametric oscillator. 44K—Satisfactory for parametric oscillator.

ACKNOWLEDGMENT

The authors would like to thank Miss D. M. Dodd for infrared measurements, A. M. Glass for the computer program used in correcting the absorption measurements for reflectivity and for helpful discussions, and C. K. N. Patel for carrier concentration measurements. The authors are especially grateful to R. C. Keezer for several discussions, to J. H. Wernick for the use of the growth apparatus and H. J. Levenstein and C. D. Capio for Laue pictures.

REFERENCES

1. PATEL, C. K. N., *Phys. Rev.* **136**, A 1187 (1964).
2. CALDWELL, R. S. and FAN, Y., *Phys. Rev.* **114**, 664 (1959).
3. PATEL, C. K. N., *Phys. Rev. Letters* **15**, 1027 (1965).

4. PATEL, C. K. N., *Appl. Phys. Letters* **9**, 332 (1966).
5. BATTOM, V., *Science* **115**, 570 (1952).
6. RIGAUX, C., private communication.
7. WEIDEL, J., *Z. Naturf.* **9a**, 697 (1954).
8. DAVIES, T., *J. Appl. Phys.* **28**, 1217 (1957).
9. KEEZER, R. C., *Tellurium Crystal Growth*, University of Michigan, U.S. Army Electronics Command, Contract DA-36-039-SC78801, December 1963.
10. SUITS, G., *Proc. IRIS*, **3**, 105 (1955).
11. KEEZER, R. C., private communication.

DISCUSSION

BECKER: Were the criteria for parametric amplification due to differences in free carrier absorption or crystal imperfections?

LAUDISE: Crystals were found not to be useful for parametric experiments whenever the carrier concentration was much above $10^{15}/cm^3$ and the perfection was poor as indicated by Schulz or Laue pictures or by examination of the external morphology.

WOOD: Were there any indications of impurity compensation in your samples of lowest carrier concentration?

LAUDISE: Our carrier concentrations were determined only by means of thermal electric measurements at room temperature. Without resistivity and mobility measurements as a function of temperature, we are unable to decide whether impurity compensation is important.

DUFRESNE: We are currently growing large single crystals of tellurium by the Czochralski method at the Noranda Research Centre under conditions similar to yours. We have observed the presence of depressions in the three smaller faces of some of the crystals. The depressions are 1–5 mm across and 1–3 cm long and have not been related to the growth conditions. We have observed that lineage is present in the depressions with grain rotation of 2–10°. This lineage penetrated the otherwise single crystals by approximately 2 mm. Did you observe such defects and can you suggest any explanation as to their origin?

LAUDISE: We have not observed any such defects, and I doubt whether they are present in our crystals. We have taken Schulz† pictures and have not observed lineage of this sort.

† *Comment.* In both selenium and tellurium the scale of imperfections is too subtle to be detected with Laue pictures (at least in the best crystals) and, given our present level of understanding, too difficult to interpret on the dislocation etch pit level. Therefore, I would suggest that the Schulz technique,‡ which is capable of resolving very low angle grain boundaries, could be applied with great profit to these materials. However, I do not mean by this that once we have passed through the level of understanding perfections on the low angle grain boundary level, we should not be interested in perfections on the dislocation level. These will clearly be of great importance given the parallel between the expected structure of selenium and many other organic polymeric materials which have been studied.

‡ Schulz, L. G., *Trans. AIME*, **200**, 1082 (1954).

† Shulz, L. G., *Trans. AIME* **200**, 1082 (1954).

COORDINATION AND THERMAL MOTION IN CRYSTALLINE SELENIUM AND TELLURIUM

P. UNGER and P. CHERIN

Research and Engineering Center, Xerox Corporation, Webster, N.Y.

THE crystal structures of three known crystalline phases of selenium and one of tellurium were determined by X-ray diffraction techniques on single crystals.

The space groups and lattice parameters are shown in Table 1. The trigonal forms of selenium and tellurium have space groups $P3_221$ or $P3_121$ depending on the direction of the screw axis along the chains. The lattice parameters

TABLE 1. *Unit Cell Dimensions*

Phase	S.G.	a (Å)	b (Å)	c (Å)	β (or γ)
Trigonal Te	$P3_121$	4.4572		5.9290	120°
Trigonal Se	$P3_121$	4.3662		4.9536	120°
α-Monoclinic Se	$P2_1/n$	9.054	9.083	11.601	90.81°
β-Monoclinic Se	$P2_1/a$	12.85	8.07	9.31	93.13°

given for trigonal selenium and tellurium were done by Swanson, Gilfrich and Urgrinic[1] and Swanson and Tate[2] respectively. The unit cell has three molecules lying on a helical chain of symmetry 3_1 or 3_2. These chains pass through the corners of a hexagonal lattice. The atoms lie on special positions on a two-fold rotation axis so that only one position parameter, x, needed to be determined.

Two-dimensional hk0 data were collected using the stationary crystal–stationary counter method on a General Electric goniostat with CuKα and MoKα filtered radiation. In each case a needle crystal of hexagonal cross-section was mounted along the needle (c) axis. Absorption corrections were made assuming the needles to have been cylindrical. To reduce errors caused by this approximation a full set of equivalent reflections were taken and averaged. Least squares refinement was done starting with Bradley's parameters.[3] In selenium convergence was rapid to a reliability of 3 per cent where reliability is defined as

$$R = \sum_{i=1}^{Nobs} |F_{obs} - F_{cal}| \Big/ \sum_{i=1}^{Nobs} |F_{obs}|$$

The results are in Table 2. The Se–Se bond length of 2.373 Å is longer than that reported previously[3] and thermal motion is rather isotropic normal to the c-axis. Since only two-dimensional data were taken, motion along c could not be determined.

TABLE 2. *Thermal and Position Parameters of Trigonal Selenium*

	Cherin and Unger (1967)	Bradley (1924)
x	0.2254 ± 0.0009	0.217
Bond length (Å)	2.373 ± 0.0055	2.328
Bond angle	103.1 ± 0.20°	104.75°
r.m.s. (along a) (Å)	0.150	
r.m.s. (perpendicular ac)(Å)	0.174	
R	3.2 %	

Figure 1 shows the trigonal selenium structure. The two nearest atoms are on the chain at a distance 2.373 Å bonding at an angle of 103.1° and the four next nearest are at 3.436 Å on neighboring chains. On the same chain second nearest neighbors are at 3.716 Å.

The tellurium data were refined to $R = 5$ per cent. The bond length shown in Table 3 is almost 0.5 Å larger than that for trigonal selenium, and slightly

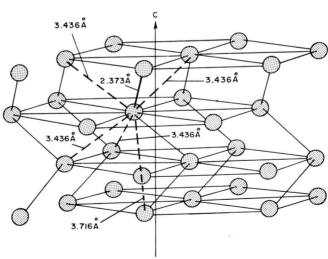

FIG. 1. Trigonal selenium.

TABLE 3. *Thermal and Position Parameters of Trigonal Tellurium*

	Cherin and Unger (1967)	Bradley (1924)
x	0.2633 ± 0.005	0.269
Bond length (Å)	2.835 ± 0.002	2.864
Bond angle	103.2 ± 0.1°	102.2°
r.m.s. (along a) (Å)	0.172 ± 0.002	
r.m.s. (perpendicular to ac) (Å)	0.167 ± 0.002	
R	5.0	

smaller than Bradley's value.[3] The bond angle of tellurium is very close to that for selenium and larger than Bradley's value.[3] The thermal motion is similar in magnitude to that found in selenium and also almost isotropic in the plane normal to c. Figure 2 shows the tellurium structure. The bond length is 2.835 Å. The next nearest neighbor distance is 3.49 Å which is very close to that for selenium even though the bond length has greatly increased. The next nearest distance on the same chain is 4.45 Å.

Selenium also crystallizes in two monoclinic phases called α- and β-monoclinic selenium. The space group and lattice parameters are given in Table 1. For α-monoclinic selenium the lattice parameters were refined by a least squares method to fit observed powder diffractometry data. Burbank's β-monoclinic selenium lattice parameters are used. Integrating Weissenburg film techniques and microdensitometry were used to collect α-monoclinic selenium intensity data. For β-monoclinic selenium automatic diffractometer

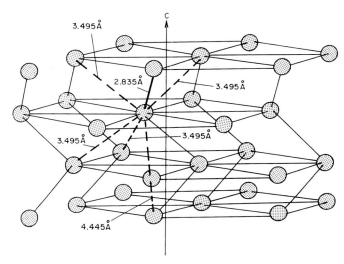

FIG. 2. Trigonal tellurium.

TABLE 4. *Bond Lengths in α-Monoclinic Selenium*

	Bond distances (Å)			Bond angles	
	Cherin and Unger (1966)	Burbank (1951)		Cherin and Unger (1966)	Burbank (1951)
d_{12}	2.325	2.325	α_1	106.7	107.7
d_{25}	2.313	2.313	α_2	107.2	106.3
d_{34}	2.331	2.356	α_3	108.8	109.2
d_{45}	2.313	2.322	α_4	104.8	101.7
d_{56}	2.307	2.364	α_5	107.7	107.4
d_{67}	2.360	2.340	α_6	105.2	103.7
d_{78}	2.311	2.352	α_7	103.1	102.8
d_{81}	2.321	2.308	α_8	103.5	104.1
d	2.318±0.007	2.335±0.018 Å	$\bar{\alpha}$	105.9±1.7	105.4±2.3
R	9.4 %				

methods were used in addition to the above film techniques. The α-monoclinic selenium crystal used was a many-faceted polyhedron and was approximated by a sphere for absorption corrections. With Burbank's[4] results as starting parameters the α-monoclinic selenium data were refined to 9 per cent by least squares analysis using anisotropic temperature factors. The bond lengths in the eight-membered ring molecule, given in Table 4, are shorter than those

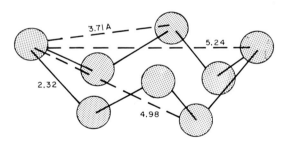

FIG. 3. Interatomic distances in α-selenium.

reported by Burbank[4] and considerably shorter than the length determined for trigonal selenium. The bond angle is slightly larger than that for trigonal selenium. Thermal motion was found to be fairly isotropic. When the molecule was treated as a rigid body, root mean square amplitudes of vibration along the three major axes ranged from 0.14 to 0.20 Å and 2–3° rotations about axes through the center. Intermolecular distances start at 3.58 Å which is shorter than the Van der Waals Se–Se distance of 4.0 Å. However, there is one shorter intermolecular distance of 3.48 Å. Figure 3 shows the interatomic distances in α-monoclinic selenium.

In the case of β-monoclinic selenium an eight-membered ring[5] and a curved open chain[6] have been proposed. The curved chain model was tried as a starting point for the three-dimensional data in the present study. The least squares refinement would not converge and R remained above 30 per cent. Using the eight-membered ring model, our film data converged to $R=17$ per cent. The automatic diffractometer data converged to $R=9$ per cent with isotropic temperature factors. Figure 4 shows the Se–Se average interatomic

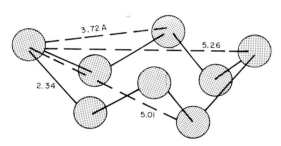

FIG. 4. Interatomic distances in β-selenium.

lengths for the molecule. These lengths are slightly longer than in α-monoclinic selenium. The difference of 0.02 Å in the Se–Se bond length is just at the limit of error and may be significant. The average bond angle is 105.5°, which is essentially the same as in α-monoclinic selenium. Because the bond distance and next nearest distance in the molecules of the three forms of selenium are so similar, distinguishing rings from chains in amorphous selenium using radial distribution is difficult.

Thermal motion in β-monoclinic selenium was not studied because the crystal was not close enough to a cylinder for this simple absorption correction to be a good approximation. The absorption approximation will not have a significant effect on the position parameters. The thermal motion is probably not isotropic as it is for α-monoclinic selenium because one atom in the β-monoclinic selenium ring approaches two other rings much more closely than in the case of α-monoclinic selenium.

Both α- and β-monoclinic selenium have neighbors on different molecules 3.58 Å away but in β-monoclinic selenium the nearest approach of atoms in adjacent molecules is only 3.44 Å. This is very close to that found in trigonal selenium. In fact the configurations of five atoms shown in Fig. 5 are very similar in these two forms of selenium. Very few measurements of the physical properties of β-monoclinic selenium have been made. However, on the basis of the intermolecular distances it is expected that the conductivity in β-monoclinic selenium will be similar to that of trigonal selenium in the (111) plane.

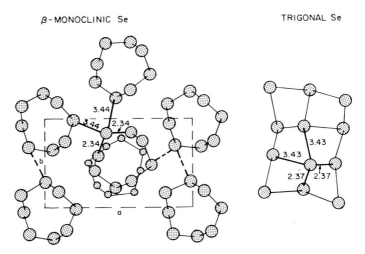

Fig. 5. β-monoclinic selenium. Trigonal selenium.

The differences in the stacking of molecules in α- and β-monoclinic selenium are shown in Fig. 6. Both drawings were done almost normal to the plane of the molecules. In β-monoclinic selenium the molecules stack parallel to each other. In α-monoclinic selenium there are two stacking directions.

While choosing β-monoclinic selenium crystals for study, single-crystal data were taken with a crystal with morphology very much like β-monoclinic selenium. However, it was clear from the Weissenburg measurements that this new crystal, which was subsequently called β'-monoclinic selenium, was not identical with the crystal of β-monoclinic selenium. The most intense

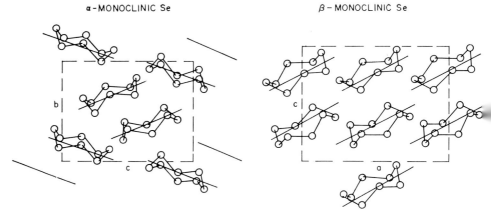

Fig. 6. α-monoclinic selenium. β-monoclinic selenium.

reflections of the a^*b^* net of β'-monoclinic selenium directly superimposed upon those for β-monoclinic selenium, but in each photograph there were reflections not on the other. The β'-monoclinic selenium crystal required a b^*-axis one-third the size of the b^* of β-monoclinic selenium in order to index all its reflections. The cell derived from this is base centered and is shown in Table 5. However, since no symmetry was observed, the unit cell is triclinic and the reduced dimensions are given in Table 5. The space group is either P1 or P$\bar{1}$. We believe this is a new form of selenium similar to β-monoclinic selenium.

TABLE 5. Unit Cell Dimensions of Monoclinic Selenium

Phase	S.G.	a (Å)	b (Å)	c (Å)	α	β	γ
β-Monoclinic selenium	P2$_1$/a	12.85	8.07	9.31	90°	93.13°	90°
β'-Monoclinic selenium	P1	12.82	13.99	9.43	93°	90°	116°
β'-Monoclinic selenium (base centered)		12.82	24.88	9.43	90°	90°	99°

REFERENCES

1. SWANSON, H. E., GILFRICH, N. T. and UGRINIC, G. M., *NBS Circular* **539**, Vol. V (1955).
2. SWANSON, H. E. and TATGE, E., *NBS Report* **2202** (1951).
3. BRADLEY, A. J., *Phil. Mag.* **48**, 477 (1924).
4. BURBANK, R. D., *Acta Cryst.* **4**, 140 (1951).
5. MARSH, R. E., PAULING, L. and McCULLOUGH, J. D., *Acta Cryst.* **6**, 71 (1953).
6. BURBANK, R. D., *Acta Cryst.* **5**, 236 (1952).

OPTICAL PROPERTIES

INFRARED-ACTIVE LATTICE VIBRATIONS IN AMORPHOUS SELENIUM

A. TAUSEND

II. Physikalisches Institut der Technischen Universität Berlin, Berlin, Germany

INTRODUCTION

Infrared spectroscopy is a very important tool for the investigation of semiconductors and has been applied successfully by several authors to a study of the optical behavior of amorphous selenium. These studies have been facilitated by the high transparence of selenium in the infrared region and by the relative ease of preparing samples with good optical surfaces.

The results of some authors are in good agreement, but in other cases there are serious discrepancies. It is felt that this situation is caused by impurities in the selenium and by the method of sample preparation. The purpose of the present work was to measure the optical constants of extremely pure and doped amorphous selenium.

PREPARATION OF THE SAMPLES

The material used had a purity of 99.995 per cent. According to Vǎsko[1] selenium can be purified from gases when heated. The purity can be further improved by distillation in a high vacuum. In this study the degree of purification was controlled by radiochemical methods.

Selenium can be produced by three methods:

1. Oxidation of the 99.7 per cent selenium to SeO_2 and reduction by sulfur dioxide.
2. Formation of Na_2Se and subsequent reduction.
3. Thermal dissociation of H_2Se. This selenium is extremely pure, but can be obtained in small quantities only. In this method the impurity concentrations of the investigated elements can be held smaller by a factor of more than 10 compared with other methods of purification.

For optical investigations the samples were prepared in the following manner. Liquid selenium was placed between two thin quartz plates. By mounting one quartz plate on a copper disc it was possible to quench the sample rapidly. The samples had optically good surfaces.

RESULTS AND DISCUSSION

Optical data were determined for selenium of various origins and preparation methods (Fig. 1). In the upper part of Fig. 1 the transmission is shown for three samples with different thicknesses and a purity of 99.995 per cent. In addition to the strong absorption bands at 13.7 and 20.3 microns there are some weaker bands. The two strong bands are caused by the pure selenium lattice concluded from the similarity of the infrared spectra of selenium and sulfur. In both spectra the same band arrangement is found, whereas the energy position is different because of the different atomic weights. These bands do not result from fundamental vibrations.

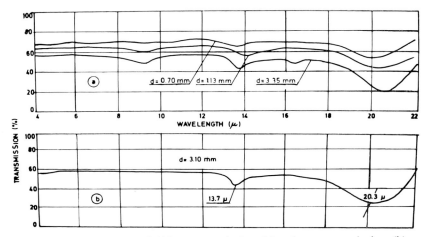

FIG. 1. Transmission of 99.995 per cent selenium (a) and very pure selenium (b).

One of the most important aims of the present study was to determine whether the weak bands are caused by selenium itself or by impurities. In the lower part of Fig. 1 the transmission is shown for extremely pure selenium. Besides the strong bands there are no further ones, so that the small bands in the upper part of Fig. 1 are due to impurities.

The experiments at low temperatures demonstrate that the position and shape of the two bands are relatively independent of temperature (Fig. 2). The small variations observed may result from the experimental set-up. This fact means that there is a 1-phonon process. This 1-phonon process results from dipole moments of the first order. This is an indication that the 13.7- and 20.3-micron bands are caused by the undisturbed selenium lattice and that the vibrations are predominantly limited to the chains.

Geick[2] has shown by theoretical group methods that the low symmetry of the selenium molecule allows absorption processes which are forbidden if the lattice is of high symmetry and the binding is strongly homopolar.

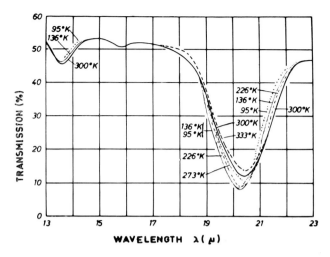

FIG. 2. Temperature dependence of the 13.7- and 20.3-micron bands.

Chlorine-doped selenium shows a broad absorption band at 10.6 microns (Fig. 3). This band allows some conclusions on the binding of the chlorine atom in selenium.

Selenium has a strong chemical affinity for chlorine. The band just mentioned is probably caused by one of the two possible selenium–chlorine compounds (Se_2Cl_2 and $SeCl_4$). Excess selenium favors the formation of Se_2Cl_2. If selenium is doped with chlorine, it can be expected that it is bound in the form of Se_2Cl_2. The best proof would be to compare the spectra of Se_2Cl_2 and

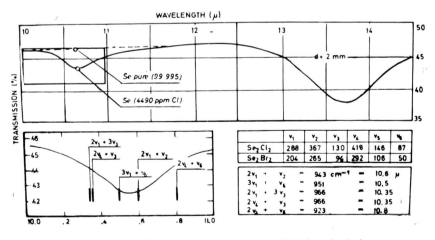

FIG. 3. Infrared and Raman spectra of pure and Cl-doped selenium.

chlorine-doped selenium. Unfortunately it is difficult to prepare Se_2Cl_2 because there always is a tendency towards the formation of $SeCl_4$. Stammreich and Forneris[3] measured the six fundamental vibrations in the Raman spectrum of Se_2Cl_2. A comparison of these results with the present infrared investigations shows that the 10.6-micron band may be explained as a combination of some upper vibrations of the Se_2Cl_2 molecule. Some combinations give the values which are shown in the lower left part of Fig. 3. This may be an explanation of the broad chlorine band at 10.6 microns. A combination of the analogous fundamental vibrations of the Se_2Br_2 molecules also leads to a band at

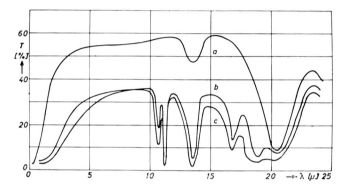

FIG. 4. Transmission vs. wavelength for pure and oxygen-doped selenium (a) pure, (b) 500 ppm, $d=1.95$ mm, (c) 500 ppm, $d=2.7$ mm.

15 microns. The 10.6-micron band is only caused by the compound Se_2Cl_2. Some physical properties of amorphous chlorine-doped selenium may be influenced by other types of bonding which are not infrared-sensitive.

Figure 4 shows the spectra of two 500 ppm oxygen-doped samples and the spectrum of pure selenium. As reported earlier oxygen causes some further characteristic bands. These bands are only found if oxygen is chemically bound to selenium like in solid SeO_2. No influence on the infrared spectrum is obtained by passing oxygen through liquid selenium.

In Fig. 5 oxygen-doped selenium is compared with the spectrum of pure SeO_2. The good agreement of the two spectra provides an indication of the manner in which oxygen is combined in oxygen-doped selenium.

According to X-ray investigations of McCullough[4] the SeO_2 lattice consists of long chains of alternating oxygen and selenium atoms at a distance of 1.78 Å. Each selenium atom is bound by a double bond to an oxygen atom. The distance between these atoms is 1.73 Å. This means that only 20–25 per cent of this bonding has the character of a pure double bond. This structure is confirmed by the Raman spectrum measured by Gerding.[5] Thus, the strongest band at 11.1 microns is associated with a valence vibration of the

Se–O sidegroup.[6] The bands between 14.4 and 19.7 microns are caused by valence vibrations in the chain. Because of the weakness of the corresponding Raman bands, the band at 14.3 microns must result from an asymmetrical valence vibration of the O–Se–O entity. The existence of some frequencies near 11 microns (valence vibration of the sidegroup Se=O) and the valence vibration of the chain are explained by the strong coupling of the structural units.

It was shown by experiments that the oxygen bands decrease very quickly if selenium is heated in air for a short time. The quantity of oxygen escaping per time unit is larger than the quantity measured by the infrared spectrum. This means that there also is dissolved oxygen, not chemically bound to selenium, which has a high mobility of diffusion.

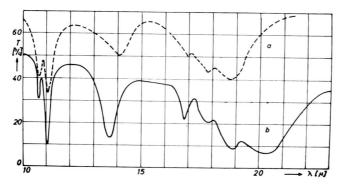

FIG. 5. Transmission vs. wavelength for (a) SeO_2, (b) oxygen-doped selenium.

Some transmission spectra of sodium-doped samples are shown in Fig. 6. Even sodium-doping of only 10 ppm causes clear bands. Selenium produced from Na_2Se often shows these bands contrary to selenium produced by other methods. These infrared measurements do not allow a final conclusion about the character of the binding. From thermal analysis the possible compounds are Na_2Se, Na_2Se_2, Na_2Se_3, Na_2Se_4 and Na_2Se_6. The absorption is non-linear with sodium concentration. Sodium has a small distribution coefficient and the clusters influence very strongly the crystallization properties in small domains. Therefore, the total sodium concentration is not particularly significant.

Thallium-doped samples (500 ppm) show very weak bands at 9.4 and 16.6 microns. These bands probably result from thallium selenide. Small concentrations of thallium in selenium are spectroscopically inactive, because the thallium appears as Tl^{+3} ion. A free ion may influence the structure of chains or rings but it cannot be detected by infrared spectrometry.

Finally, a remark should be made about the absorption bands at about 3 microns which have been found in this laboratory in hexagonal selenium

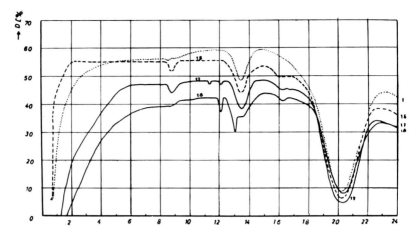

Fig. 6. Transmission vs. wavelength. 1, Pure selenium. 15, 10 ppm Na. 17, 100 ppm Na. 18, 500 ppm Na. $d=2.8$ mm.

single crystals. Lanyon[7] believed that these bands are caused by a tellurium impurity. Owing to the oscillator strength of these bands the tellurium concentration must be in the order of 10^3 ppm. For verification the infrared spectra of amorphous selenium samples doped with tellurium up to concentrations of 10^3 ppm were measured. No bands near 3 microns were found. Kessler and Sutter[8] have shown by other methods that bands at 3 microns only result from transitions in the undoped selenium.

REFERENCES

1. Vaško, A., private communication.
2. Geick, R., private communication.
3. Stammreich, H. and Forneris, R., *Spectrochim. Acta* **8**, 46 (1956).
4. McCullough, J. D., *J. Am. Chem. Soc.* **59**, 789 (1937).
5. Gerding, H., *Rec. Trav. Chim.* **60**, 728 (1941).
6. Giguere, P. A. and Falk, M., *Spectrochim. Acta* **16**, 1 (1960).
7. Lanyon, H. P. D., *J. Appl. Phys.* **35**, 1516 (1964).
8. Kessler, F. R. and Sutter, E., *Z. Physik* **173**, 54 (1963).

DISCUSSION

Queisser: Am I correct in understanding that you interpret all infrared absorption bands as being due to lattice vibrations? Is there any evidence for electron transitions involving impurities?

Tausend: We attribute the absorption bands in the region near 3 microns to electron transitions in the $4p$ valence bands. Kessler and Sutter were able to confirm this by two convincing experiments:

(a) They measured the time dependence of an infrared light beam in transmission during the irradiation of the surface by a pulsed light source.
(b) The authors also found these bands in photoemission. These results cannot be explained by lattice vibrations.

GÜNTHER: Could your method differentiate between Se_2Cl_2 and a structure containing Se-chains capped by the halogen Se_xCl_2. The same type of structure could apply to oxygen-containing impurities. In addition a side chain attachment like Se–Se–Se should be possible.

TAUSEND: We are able to differentiate between Se_2Cl_2 and a Se-chain-like structure such as Se_xCl_2 but only for infrared- and Raman-active lattice vibrations. Therefore, we compared the spectra of pure Se_2Cl_2 and chlorine-doped selenium. We found good agreement between the spectra of Se_2Cl_2 and chlorine-doped selenium so that we have to assume that the configuration Se_2Cl_2 is dominant. Corresponding to the formula

$$\omega = \frac{1}{2\pi}\sqrt{\left(k\left[\frac{1}{nM_{Se}} + \frac{1}{M_{Cl}}\right]\right)}$$

$n = 1, 2, 3, \ldots$ $\qquad M_{Se}$ = mass of a Se atom
k = constant $\qquad M_{Cl}$ = mass of a Cl atom

one should expect a shift of the frequencies for increasing n in comparison with the spectra of Se_2Cl_2. We did not find this shift. Concerning oxygen, two types of binding are possible as shown below:

[Diagram showing Se-O structures with bond lengths 1.78 Å and 1.73 Å]

LANYON: In your paper you commented on my suggestion at the London conference that the 3.4-micron band in crystalline selenium may be associated with tellurium impurities. You said that you could not measure any effect up to 10^3 ppm in the amorphous phase. I did not see bands of this type developing even up to 50 atomic per cent thallium incorporated into the amorphous phase. However, the band did show up very clearly in the crystalline phase. My comments at the London conference were only referring to this crystalline phase, not to the amorphous phase.

TAUSEND: We reported on the 3-micron bands for the first time in 1961 and attributed these bands to electron transitions. The transmission is in agreement with Beer's Law which means that we can exclude surface effects. Kessler and Sutter found these bands in the amorphous modification, too. Here the bands are so weak that we were not able to detect them in the transmission by means of our equipment. Recently these bands were also found in photoemission.

LUCOVSKY: Spectroscopically one can distinguish between compounds such as H_2S and H_2S_x, $x \gg 1$, i.e. H–S–H; H–S–S...S–S–H. The same would apply for Cl_2Se and Cl_2Se_x, $x \gg 1$.

You stated that the 490 cm^{-1} (20.3-micron) band is a chain fundamental. Could you clarify this statement?

TAUSEND: The 20.3-micron band is caused by the pure selenium chain and not by impurities. Here we have a combination vibration.

OPTICAL PROPERTIES OF AMORPHOUS AND LIQUID SELENIUM

A. VAŠKO

Institute of Radio Engineering and Electronics, Czechoslovak Academy of Sciences, Prague, Czechoslovakia

It is well known that since their discovery, defects in classical optical glasses have been systematically investigated and optical glass itself is characterized as a glass free from defects. However, practically no attention has been paid so far to the investigation of these defects in the non-classical glasses such as amorphous selenium and glassy semiconductors, despite the fact that the defects are manifest here more strongly.

Therefore, a systematic investigation of the internal defects in these glasses was undertaken using infrared techniques. The studies have shown that the samples of these glassy materials, prepared by a technology currently described in the literature, exhibit considerable internal defects which in some cases differ in their intensity from sample to sample so much that the material cannot be considered as a defined material at all.

For this investigation the following methods have been used: infrared microscopy in normal or polarized radiation with normal or dark field; infrared polarimetry and the "Schlieren method" for the infrared region. A simplified optical arrangement of the Toepler–Schlieren method which was modified for the infrared region is shown in Fig. 1. The incandescent lamp L, in the

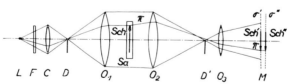

Fig. 1. Optical arrangement of the modified Toepler–Schlieren method for the infrared:

L	incandescent lamp source of radiation	D'	Schlieren-diaphragm
F	infrared filter	Sa	sample to be investigated
C	condenser	Sch	Schliere in the sample
D	secondary source of infrared radiation	Sch'	infrared image of the Schliere
		Sch"	visible image of the Schliere
$O_{1,2,3}$	objectives	M	image converter

For more details the reader is referred to the monograph: A. Vaško, *Infrared Radiation*, 1968, Iliffe Books, Ltd., London.

simplest case, or for a better resolution and contrast, a laser source, is used as a primary source of radiation. As can be seen from the passage of rays an infrared image Sch′ of the Schliere, Sch, is formed in the plane σ′, which can be then converted into a visible image Sch″ by one of the methods M for making an infrared image visible.

As an example infrared pictures of defects in amorphous or liquid selenium both pure and doped are shown in Fig. 2. This figure shows internal stresses and deformations of the surface of a cast glassy selenium plate, taken by the Schlieren method. The diameter of this picture is 30 mm.

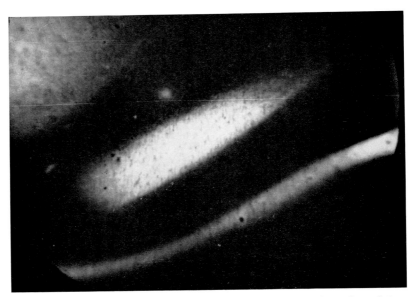

FIG. 2. Infrared picture of internal stresses, Schlieren and deformation of the surface of a sample of cast glassy selenium plate. The diameter of the picture is 30 mm.

Very frequent defects in samples of glassy selenium are the air bubbles and cavities. Figure 3 shows an infrared micrograph of air bubbles in the mass of glassy selenium taken in transmitted light, and Fig. 4 shows the same object micrographed in an "illumination with half dark field". Various spatial effects can be achieved by a suitable choice of the angle under which the "illuminating" radiation is incident on the investigated sample. This illumination can form an arbitrary angle between the usual direction used for the examination in transmitted light and that used for work with reflected light. The latter technique substantially facilitates the identification of the defects. It is interesting to note that Fig. 4 was taken on a sample on which formerly many fundamental measurements of physical properties had been made in a well known selenium laboratory. Figure 5 is an infrared picture of the porous structure in the axial part of the ingot of glassy selenium, taken in polarized infrared radiation.

Still more defects can be found in samples of doped glassy selenium. Figure 6 is an infrared picture of residual concentration Schlieren in selenium doped

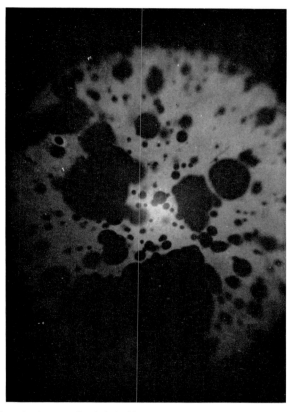

FIG. 3. Infrared micrograph of air bubbles in the mass of glassy selenium taken in transmitted light, magnification 15×.

with 20 atomic per cent germanium despite precautions taken to obtain homogeneous ingots. Figure 7 shows an infrared micrograph of another sample of selenium doped with 20 atomic per cent germanium, with a partially segregated second phase, probably $GeSe_2$. The ingot was prepared by current technology and then cooled slowly to room temperature. As the magnification of the electron lens of the image converter increases with the radius, the 10-micron divisions of the scale increase from the center to the periphery of the Figure. Figure 8 shows the same defect in a sample of selenium doped with

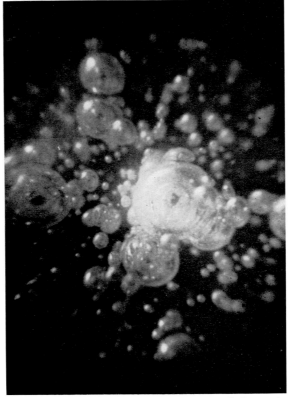

Fig. 4. Infrared micrograph of air bubbles in the mass of glassy selenium in an "illumination with half dark field", magnification 15×.

5 per cent arsenic. Figure 9 shows intensive Schlieren despite prolonged stirring, in liquid selenium doped with sulfur taken by the Schlieren method.

The correlations between the average patterns of the defects and the corresponding transmissivity will be considered. Figure 10 shows infrared images of average patterns of defects found in five blocks of amorphous selenium (nos. 1 to 5) of equal thickness (15 mm) and containing varying concentrations of air bubbles and mechanical admixtures respectively, viz.:

> Image no. 1 is that of a sample containing a high concentration of large air bubbles. Its transmissivity at 2 microns was 12.5 per cent.
> Image no. 2 is that of a sample containing a high concentration of large mechanical admixtures. Its transmissivity at 2 microns was 23 per cent.
> Images nos. 3, 4 and 5 are those of samples containing rapidly decreasing concentrations of air bubbles. Their transmissivities at 2 microns were 44, 61 and 62 per cent, respectively

Image no. 6 shows the screen of the image converter itself. (Some small defects on the screen are shown.) All the pictures of defect patterns are infrared micrographs having a width of 1 mm.

In all the cases shown the infrared images were photographed from the screen of an infrared image converter. Infrared images were formed by wavelengths in the range centered around 1 micron.

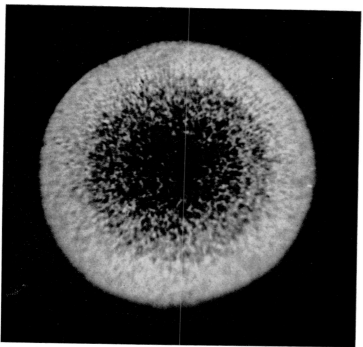

FIG. 5. Infrared picture of the porous structure in the axial part of the ingot of glassy selenium made in polarized infrared radiation. The diameter of the ingot is 20 mm.

With the aid of the foregoing techniques, one can also detect with great sensitivity the mechanical admixtures, the first traces of the crystalline phase, the frozen-in states corresponding to higher temperatures, and gradients of the index of refraction.

The absorption coefficient of amorphous selenium was measured on selected samples, prepared from pure selenium 99.9999 + per cent, completely free from defects. Figure 11 shows measurement at the absorption edge.

The absorption coefficient was calculated from the measured transmissivity and reflectivity of a thick layer according to the exact formula which is valid for multiple reflections shown in Fig. 11.

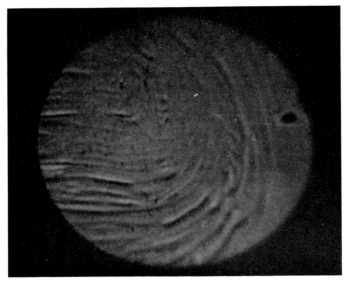

Fig. 6. Residual concentration Schlieren in a glassy selenium doped with 20 atomic per cent germanium after prolonged homogenization of the melt.

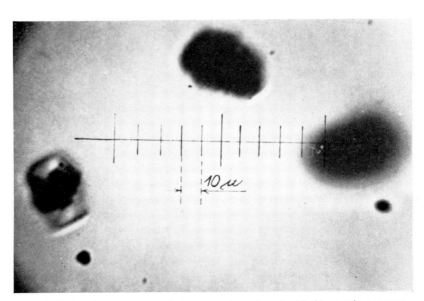

Fig. 7. Infrared micrograph of glassy selenium doped with 20 atomic per cent germanium with partially segregated second phase, probably $GeSe_2$. The ingot was prepared by current technology and cooled slowly to room temperature.

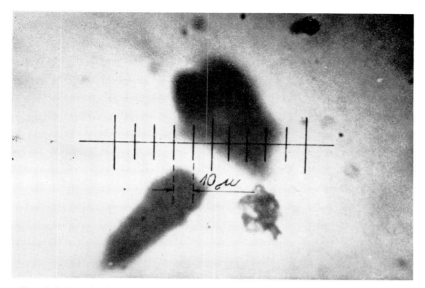

FIG. 8. Infrared micrograph of cast glassy selenium doped with 5 atomic per cent arsenic with partially segregated second phase.

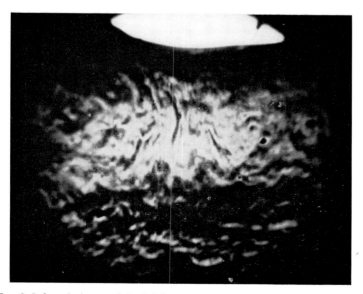

FIG. 9. Infrared picture of the Schlieren in liquid selenium doped with sulfur.

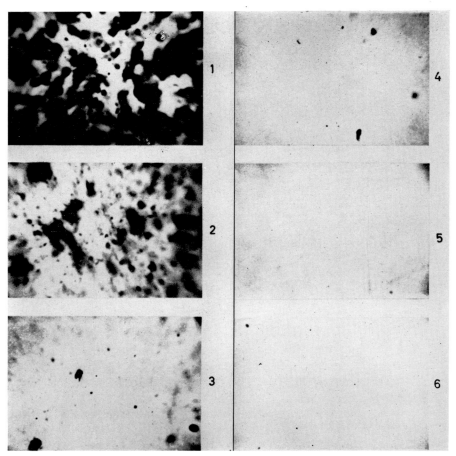

FIG. 10. Infrared images of average patterns of defects found in five blocks of amorphous selenium (Nos. 1–5) of equal thickness (15 mm) and containing varying concentrations of air bubbles and mechanical admixtures respectively. Image No. 6 shows the screen of the image converter itself.

Curve 3 is the lower boundary of the values of the absorption coefficient; the dashed part of this curve is not yet definitive.

It can be seen that the values obtained are by one to two orders of magnitude smaller, sometimes even more, than the values published by Gobrecht and Tausend[1] (curves 1, 1′) and Caldwell and Fan[2] (curve 2).

The measurements from the present study in the region from 2 to 25 microns have also shown that a substantial part of the ground "absorption continuum" of both pure and doped amorphous selenium in the infrared region, often mentioned in the literature,[3,4] is caused by the defects in the

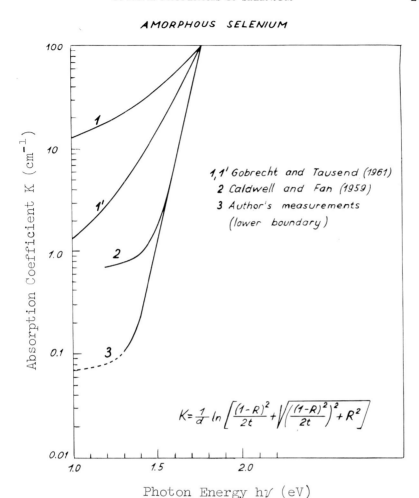

FIG. 11. The absorption coefficient of amorphous selenium at the absorption edge:

1, 1'   spread of values after Gobrecht and Tausend.[1]
2   measurements of Caldwell and Fan.[2]
3   lower boundary of the author's measurements on selected defect-free samples; the dashed part of this curve is not yet definitive.

samples. The spread of the measured values was decreased by about one order of magnitude in comparison with literature values.

It is clear that the pre-examination of defects in amorphous selenium is important not only for the study of optical properties but also for the study of other physical properties.

Since the problem of the intrinsic and impurity bands of amorphous selenium in the infrared is the subject of papers presented by Tausend and Lucovsky at this Symposium, the discussion in this paper will be confined to the bands at 13.4 and 20.4 microns and to some impurity bands that may be related to the defects in the samples.

The question, whether these absorption bands are intrinsic or impurity bands, has been a subject of controversy. Table 1 presents a comparison of the measurements of different authors.

TABLE 1. *History of Intrinsic Absorption Bands of Amorphous Selenium in the Middle IR-Region*

Author	Wavelengths in microns					
Gebbie and Cannon (1952)	13.5	16.0	20.5			
Frerichs (1953)		?				
Lecomte (1958)			Impurity band?			
Caldwell and Fan (1959)	13.5	16.1	20.4	27.2		39.5
Gobrecht and Tausend (1961)	13.6	16.0	20.3			
Author's measurements (1961)	13.4	16.0	20.4			
Abdullayev et al. (1966)	Impurity S_2O_2		20.4			
Author's present measurements	13.4	Impurity band[a]	20.4	27.2	32.6	39.5

[a] Given by Gobrecht and Tausend.

The infrared spectra of amorphous and liquid selenium and amorphous, rhombic and liquid sulfur have been determined and compared. In the upper part of Fig. 12 there is shown the spectrum of amorphous selenium, and in the lower part the spectra of amorphous and rhombic sulfur (scale factor 1.8).

This comparison shows a quantitative similarity of infrared spectra of amorphous selenium including the bands at 13.4 and 20.4 microns with strong bands of amorphous and rhombic sulfur. The comparison of the infrared spectra of selenium and sulfur in the liquid state in the temperature range from 250 to 400°C shows the same quantitative similarity as in the solid state. It has also been found that the spectrum of liquid selenium in the same temperature range is identical with the spectrum of amorphous selenium, with only a small shift of the bands towards longer wavelengths.

This comparison shows that the dominant bands of amorphous selenium in the middle infrared region are connected with the vibrations of Se_8 rings and that the two bands discussed are intrinsic vibration bands of amorphous and liquid selenium. The same holds for amorphous sulfur.

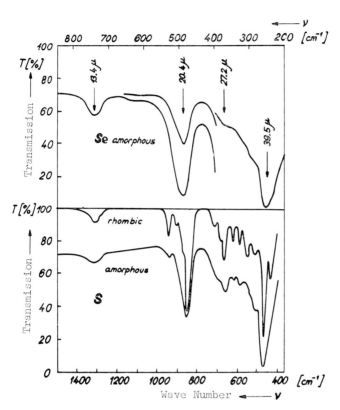

FIG. 12. A comparison of the vibration spectra of amorphous selenium and amorphous and rhombic sulfur.

As far as is known, this laboratory was the first to describe in 1964 the infrared spectrum of amorphous selenium doped with oxygen.[5] In all cases, it was found that the spectrum was identical with that of SeO_2. This SeO_2 can be removed from the selenium by outgassing. The exact wavenumbers of its five bands are indicated in Fig. 13. Note that the band at 716 cm^{-1} of SeO_2 superimposes the band of selenium at 744 cm^{-1}.

Abdullayev et al.[6] have found recently only three bands, each in reality corresponding to the super-position of two of the six bands indicated in the upper part of Fig. 13. In the samples of deoxidized selenium, as well as in selenium doped with manganese or cadmium, both of which act as a deoxidizing agent, Abdullayev did not find the three bands mentioned above. From this he concluded that the band at 13.4 microns (744 cm^{-1}) belonged to the SeO$_2$ impurity. Since after deoxidation the overall transmissivity of his samples decreased strongly, which decrease was certainly caused by the large defects in the samples resulting from the deoxidation process, the weak intrinsic band of amorphous selenium at 13.4 microns could have disappeared in the noise, even if the measurements had been carried out on thin samples only.

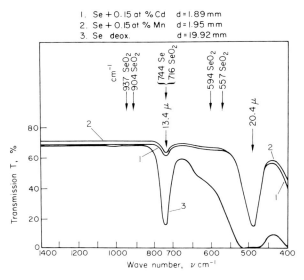

FIG. 13. Infrared spectrum of amorphous selenium doped with 0.15 atomic per cent cadmium (curve no. 1), with 0.15 atomic per cent manganese (curve no. 2) and oxygen-free selenium from Boliden (curve no. 3). The wave numbers of additional absorption bands of amorphous selenium doped with O$_2$ are marked by arrows.

Figure 13 shows the measurements from the present study both on samples prepared from high-grade pure and oxygen-free selenium kindly donated by Bolidens Gruvaktiebolag, Sweden, and samples doped with 0.15 atomic per cent manganese or 0.15 atomic per cent cadmium. A distinct band of amorphous selenium at 13.4 microns was always found as shown in Fig. 13. However, bands of SeO$_2$ have never been found. The synthesis of the samples was made in vacuum about 10^{-5} mm Hg, while great care was taken to obtain perfectly homogeneous ingots, as indicated by the high transmissivity of the samples.

In the samples of amorphous selenium doped with chlorine, in the range from 2 to 25 microns no additional band has been found that could be attributed to the doping. Similarly, doping with tellurium has yielded no absorption band. In the latter case a weak broad band has sometimes been found at 15.4 microns which is due to the presence of TeO_2 impurity. It should be pointed out that in the two cases of doping with chlorine or tellurium, the bands at 2.9 microns and 3.4 microns have not been found.

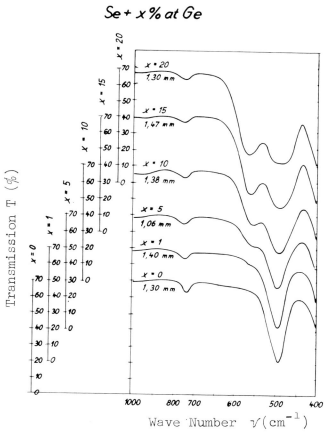

FIG. 14. Infrared spectrum of amorphous selenium pure and doped with 1 to 20 atomic per cent germanium.

Doping of selenium with germanium yields an additional band at 560 cm^{-1}. Figure 14 shows the absorption spectrum of selenium doped with 1 to 20 per cent germanium measured on defect-free samples. It has been found that the absorption coefficient of this additional band at 560 cm^{-1} is proportional to the concentration of $GeSe_2$ in the sample.

Neither the two weak absorption bands at the wavelengths of 3.4 and 9.1 microns, sometimes quoted in the literature, have been found in very pure samples of amorphous selenium free from traces of the crystalline phase.

REFERENCES

1. GOBRECHT, H. and TAUSEND, A., *Z. Physik* **161**, 205 (1961).
2. CALDWELL, R. S. and FAN, H. Y., *Phys. Rev.* **114**, 664 (1959).
3. STUKE, J., *Z. Physik* **134**, 194 (1953).
4. KAMPRATH, W., *Ann. Phys.* **9**, 382 (1962).
5. VAŠKO, A., *Phys. Status Solidi* **8**, K41 (1965).
6. ABDULLAYEV, G. B., MEKHTIEVA, S. I., ALIEV, G. M., ABDINOV, S. SH. and KERIMOVA T. G., *Phys. Status Solidi* **16**, K31 (1966).

DISCUSSION

WOOD: How were the blocks of selenium with very low absorption coefficients ($< 10^{-1}$ cm) prepared?

VAŠKO: The preparation of defect-free blocks of amorphous selenium is a very delicate operation. We have not worked out any technological procedure for this case. The preparation of these blocks depends on laboratory practice only. We have prepared these blocks by casting either in air or in vacuum.

Using the first technique, which is very simple in principle, we heated the selenium in a crucible to about 300°C and then submitted the sample several times to vacuum for degassing. Then we cast the selenium into a quartz-plated mould pre-heated to about 300°C and then let it cool rapidly to room temperature. Using this technique we were able to prepare large blocks of amorphous selenium up to $20 \times 30 \times 60$–100 mm.

In the second case, the blocks were prepared in quartz tubes about 20 mm in diameter, in vacuum (about 10^{-5} mm Hg), while great care was taken to obtain perfectly homogeneous ingots. The selenium was also pre-heated to about 300°C and thoroughly degassed.

The prepared blocks were checked by infrared techniques and only those which showed few defects (bubbles, cavities or other defects) were chosen. I must say that very few of these blocks were chosen, however.

Before grinding and optical polishing, the ingots were submitted to an annealing process at about 50°C for several hours in order to remove internal stresses.

CHAMPNESS: (1) Can you not distinguish whether an absorption band (e.g. the 13.4- and 20.4-micron bands) is due to pure selenium or to impurities by temperature shift? (2) Can amorphous selenium be used as infrared windows and lenses?

VAŠKO: (1) Yes, the proper absorption bands shift with temperature, whereas the impurity bands generally do not. (2) Amorphous selenium cannot be used as the optical material for the infrared because it becomes soft at about 40°C and tends to devitrify.

VEDAM: Can you comment on the method of preparation of the selenium glasses which show voids or air bubbles?

VAŠKO: According to our experience every block of amorphous selenium cast from the melt, which had not been thoroughly vacuum degassed before, showed more or less air bubbles or cavities. For other details reference should be made to my answer to the question of Dr. Wood.

THE STRUCTURE OF AMORPHOUS SELENIUM FROM INFRARED MEASUREMENTS

G. Lucovsky

Research and Engineering Center, Xerox Corporation, Webster, N.Y.

INTRODUCTION

The use of infrared and Raman spectroscopy as tools to study the atomic structure of oxide glasses has had only limited success. In this paper it is shown that these techniques, in particular infrared spectroscopy, are indeed more useful in helping us gain an insight into the structure of amorphous selenium. The differences in the relative degree of success are attributed to inherent differences in the nature of the respective binding forces in oxide glasses and in amorphous selenium. Oxide glasses are complex network structures which can be alternatively viewed as polymers cross-linked by strong ionic or covalent bonds, whereas amorphous selenium is believed to be a mixture of two structural species, long polymer chains (presumably helical) and eight-membered ring molecules, held together by weak Van der Waal type forces. The structural model is inferred mostly from indirect evidence, namely the presence of chains is inferred from the high viscosity of amorphous selenium at its melting point; the presence of rings from the partial solubility in CS_2; and the nature of the binding from the nature of the binding forces in the crystalline forms of selenium. Therefore, if our structure model has validity, the techniques of spectral analysis appropriate to molecular systems are indeed applicable to amorphous selenium. It is further noted that these techniques as well as those of crystal lattice dynamics cannot be readily applied to aid in the interpretation of the spectra of oxide glasses.

To characterize the vibrational modes of the ring and chain constituents, a study was made of two crystalline forms of selenium, α-monoclinic selenium and trigonal selenium,[1, 2] which are structurally related to the ring and chain components of the amorphous material. In Section II of the paper the experimental results obtained for trigonal selenium are discussed and the mode assignment is indicated. The atomic motions of these modes are discussed indicating the relative importance of inter- and intra-chain forces. In Section III experimental results for α-monoclinic selenium are presented and the vibrational mode assignment is discussed. In Section IV the infrared absorption spectrum for amorphous selenium is presented and the dominant bands are noted. Section V is concerned with the interpretation of the spectrum of amorphous selenium in terms of the vibrational modes of ring and chain components. Section VI discusses briefly infrared data for As–Se alloys, and for chlorine- and bromine-doped amorphous selenium, and gives an interpretation of the observed results.

TRIGONAL SELENIUM

The crystal lattice of trigonal selenium can be visualized as being built up of helical chains located at the corners and center of an hexagonal array. The helix within a given chain is completed every third atom. The chain or helical axis defines the c-axis of the crystal. The normal modes of the D_3 point symmetry group for a three atom molecule provide a model for the long wavelength lattice modes and as such have been discussed by Caldwell and Fan[3] and Hulin.[4] Five of the six optical modes are infrared active and five are Raman active. Four of the lattice modes are associated with two doubly degenerate type E vibrations which are both infrared active (for light polarized perpendicular to the c-axis) and Raman active. One of the modes is type A_2 and is infrared active for light polarized parallel to the c-axis. The remaining mode, type A_1, is solely Raman active. The infrared active modes are in turn split into longitudinal and transverse components by the depolarization field.

The reflectivity spectrum of trigonal selenium has been discussed in an earlier paper.[2] However, for completeness the data in Fig. 1 are presented. Strong reflection bands are observed for both principal directions of incident polarization; for light polarized parallel to the c-axis the band occurs near

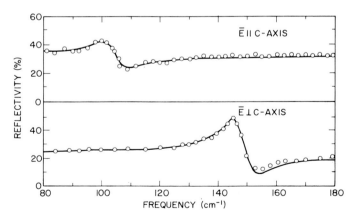

Fig. 1. Room-temperature reflectivity spectra for trigonal selenium. The solid curves are calculated using the oscillator fit parameters given in Table 1 of reference 2.

100 cm^{-1} (or 100 microns), whereas for light polarized perpendicular to the c-axis the band occurs near 140 cm^{-1} (or $\sim$70 microns). The reflectivity spectra have been analyzed using a classical harmonic oscillator formalism. For the A_2 mode ($\bar{E}_\parallel$ c-axis) the frequency of the transverse optical mode is 102 cm^{-1}; for the E mode ($\bar{E}_\perp$ c-axis) the transverse optical frequency is 144 cm^{-1}. The frequencies of the longitudinal modes, the oscillator strengths and effective

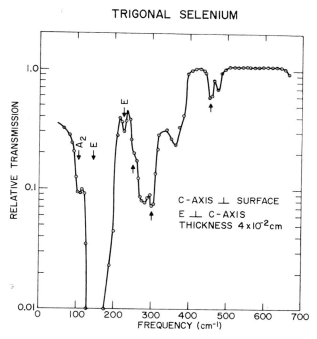

FIG. 2. Room temperature transmission spectrum for trigonal selenium for incident light polarized perpendicular to the c-axis.

charges associated with these two reststrahlen bands are discussed in detail elsewhere.[2] The third fundamental is too weak to produce a measurable reflection band, but is easily observed in transmission measurements.[1,2] Figure 2 shows the transmission spectrum for light polarized perpendicular to the c-axis. The third fundamental (type E) is associated with the sharp absorption band at about 230 cm^{-1}. Its frequency and symmetry character have been confirmed by Raman measurements.[1] The remaining structure is due to multiphonon bands. Bands marked with arrows occur only for the $\bar{E}_\perp$ c-axis polarization geometry whereas those unmarked occur for both directions of polarization.

We now focus on the vibrational motions associated with the fundamental infrared active lattice modes. Figure 3 shows the vibrational motions and indicates the frequencies of the transverse modes. For completeness the vibrational motion associated with the type A_1 Raman active vibration has also been indicated. The motion associated with the A_2 modes is a rotational motion about the helical axis. The atomic motion is all in planes perpendicular to this axis. However, the induced dipole moment is in the direction of the helical- or c-axis, the A_2 mode being infrared active for light polarized parallel to the c-axis. In the long wavelength limit the motion within each unit cell is identical, so that each of the helices is effectively rotating rigidly about its axis. The

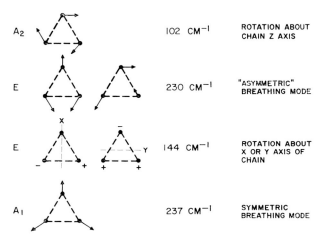

FIG. 3. Atomic motions for the long wavelength optical phonon modes of trigonal selenium.

restoring force for this motion is due to an interaction between neighboring chains; the atoms involved in this interaction are next nearest neighbors. The induced dipole moment or equivalently the macroscopic effective charge which couples the vibrational motion to the electromagnetic field is due entirely to the redistribution of charge within these next nearest neighbor bonds. This concept of a macroscopic effective charge being associated with charge redistribution is important in most covalent materials and is discussed in more detail elsewhere.[5] The most important point to note here is that the infrared activity, as well as the strength of the restoring force which in turn determines the frequency of vibration, depends very strongly on a parallel alignment of neighboring chains. To a good approximation, the higher frequency (and weaker) type E mode is due to either one of two asymmetrical breathing motions, whereas the lower frequency (and stronger) type E mode is due to either one of two rotational motions about either the "x" or "y" axes as indicated. Note that, as mentioned previously, the E modes are doubly degenerate. However, the point to be stressed here is that the restoring forces for these E modes are due to atoms within the *same* helix and that the induced macroscopic effective charge is due to charge redistribution within the bonds also of the *same* helix or chain. For completeness it is noted that the A_1 Raman active mode is also associated with the motion of atoms in a given helix and its Raman activity does not depend on the alignment of neighboring chains.

α-MONOCLINIC SELENIUM

α-Monoclinic selenium is a molecular crystal composed of eight-membered puckered ring molecules. The binding forces between the atoms in a given ring are covalent, whereas the forces holding the rings in a monoclinic lattice are of the weaker Van der Waal type. It is, therefore, expected that the infrared and Raman activity of this crystal reflects the symmetry characteristics of the molecular unit. The Se_8 molecule has symmetry D_{4d}; for this group there are at most, barring accidental degeneracies, eleven discrete vibrational frequencies. Three are associated with infrared-active modes, seven with Raman-active modes and one with a totally inactive mode. The symmetry characterization and the atomic motions of this group have been discussed by Scott and co-workers[6] in their paper dealing with the S_8 molecule. Difficulties were experienced in growing large high-quality crystals of α-monoclinic selenium. Crystals could not be polished or cut at proper orientations so that experiments using polarized light were not possible.

Figure 4 presents the fundamental absorption spectrum of α-monoclinic selenium which is dominated by three strong bands. To make a mode assignment for the observed absorption bands, the present results were compared with those of Chantry et al.[7] for S_8 in orthorhombic sulfur. It should be pointed out that the frequencies reported by Chantry et al.[7] for the bands of α-monoclinic selenium are identical to those observed for trigonal selenium and do not agree with results obtained in this laboratory for α-monoclinic selenium. It is suggested that their sample underwent conversion to the trigonal form in the course of their measurements. Paralleling the assignment from the present study of the absorption bands of Se_8 to those of S_8, the doublet at 92–97 cm^{-1} is assigned to a type E_1 vibration, the doublet at 116–122 cm^{-1} to a type B_2 vibration and the singlet at 254 cm^{-1} to a second E_1 vibration. The shoulder on the low energy side of the band at 254 cm^{-1} is assigned to a combination band. Comparing the frequencies of the corresponding modes of S_8 and Se_8, it is noted that the frequencies (f_S/f_{Se}) scale with a factor which varies from 1.87 to 2.01. Similar scaling factors are observed for corresponding pairs of Raman-active modes.[1] Figure 5 indicates a more detailed look at the entire absorption spectrum of α-monoclinic selenium in the wavelength region from 130 cm^{-1} to 650 cm^{-1}. Using the observed frequencies of the infrared and Raman modes and estimating the frequency of the inactive B_1 mode, the

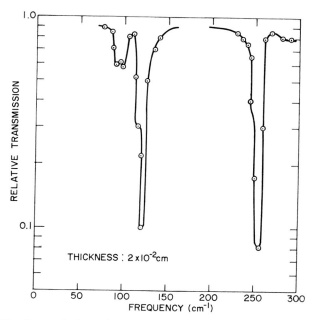

FIG. 4. Fundamental absorption spectrum of Se_8 in α-monoclinic selenium as measured at room temperature.

frequencies of the second order combination bands were computed and those occurring in the 130 to 650 cm^{-1} frequency domain in Fig. 5 have been indicated. Note that where structure is clearly resolvable, e.g. at 141 cm^{-1}, 158 cm^{-1}, 173–175 cm^{-1}, etc., the agreement between the calculated and observed frequencies is excellent. Also note that where absorption in second order is not expected, the transmission reaches a maximum, e.g. in the region from 400 to 475 cm^{-1} and beyond 550 cm^{-1}.

The spectrum of α-monoclinic selenium can be explained as being due to internal vibrations of the Se_8 molecule. The only manifestation of the crystal packing is the doublet splitting of the two low-frequency fundamentals. A similar doublet splitting has been observed in the infrared[7] and Raman[8] spectra of S_8 in orthorhombic sulfur.

AMORPHOUS SELENIUM

Transmission measurements were made on both cast and evaporated samples of amorphous selenium. Figure 6 shows the absorption spectrum of a cast sample. The dominant absorption bands occur at 95 cm^{-1}, 135 cm^{-1} and

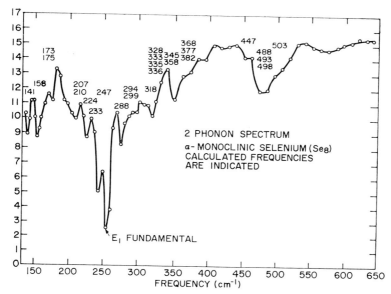

Fig. 5. Absorption spectrum at room temperature of α-monoclinic selenium showing second-order bands in the 130 to 650 cm^{-1} frequency domain.

254 cm^{-1}. These three bands also dominate the absorption spectrum of evaporated samples. Note that evaporated selenium films were either stripped from aluminum substrates, formed on transmitting polyethylene substrates or viewed in a back reflection-transmission configuration on aluminum substrates. Results were essentially independent of preparative method. Weaker structure appears as two shoulders, one at 120 cm^{-1} and one at 230 cm^{-1}, as a broad band from 280 to 380 cm^{-1}, and as weak well-resolved bands at 490 cm^{-1} and 744 cm^{-1}. The band at 744 cm^{-1} is not shown in the figure. Similar spectra for amorphous selenium have been reported by other workers.[2, 9]

The low-temperature behavior of the bands at 490 cm^{-1} and 744 cm^{-1} have been examined and a decrease in absorption strength was observed in going from room temperature to liquid nitrogen temperature. This behavior is indicative of higher-order absorption processes, e.g. second- or third-order combination bands. The temperature dependence of the rest of the spectrum has not as yet been studied.

DISCUSSION OF THE SPECTRUM OF AMORPHOUS SELENIUM

The absorption spectrum of amorphous selenium is now compared with those of α-monoclinic and trigonal selenium. The absorption spectrum bears a much closer resemblance to that of Se$_8$ molecules as observed in α-monoclinic selenium than it does to the spectrum of chains in trigonal selenium (for either direction of polarization or for an average orientation, e.g. polarized light with the electric field vector oriented 45 degrees to the *c*-axis and the propagation vector normal to the *c*-axis). Figure 7 indicates a superposition of the spectra of Se_8 in α-monoclinic selenium and amorphous selenium. The band in amorphous selenium at 95 cm^{-1} lines up with the center of the 92–97 cm^{-1}

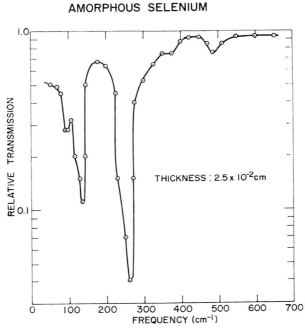

FIG. 6. Room temperature absorption spectrum of a cast sample of amorphous selenium.

doublet. Similarly the shoulder at 120 cm^{-1} lines up with the doublet at 116–122 cm^{-1} and the band at 254 cm^{-1} lines up with the singlet at 254 cm^{-1}. The bands of Se_8 in amorphous selenium are considerably broader than the corresponding bands of the Se_8 molecule in α-monoclinic selenium. The only strong structure in the spectrum of amorphous selenium that is not coincident in frequency with a fundamental band of the Se_8 molecule is the band at 135 cm^{-1}. However, the absorption here occurs quite close to the frequency of one of the *E* modes of trigonal selenium. As indicated earlier, the restoring force and effective charge for this mode are derived solely from atoms within a single helix. Considering the two other fundamental modes of trigonal selenium, no structure is observed at 102 cm^{-1} (A_2 mode) and only a very weak shoulder on the 254-cm^{-1} band near 230 cm^{-1} (second E mode). Heat treating the amorphous selenium results in partial crystallization (to the trigonal form) and induces, as expected, absorption near 102 cm^{-1}. Further, note that the weak band at 490 cm^{-1} lines up with the second-order bands of α-monoclinic selenium. The same observation is made for the broad absorption band from 280 cm^{-1} to 390 cm^{-1}. As indicated earlier, the strength of the

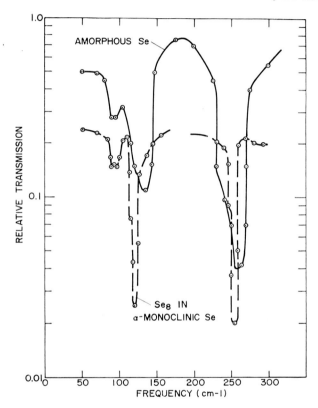

FIG. 7. Comparison of the room-temperature spectra of Se$_8$ in α-monoclinic selenium and amorphous selenium.

absorption band at 490 cm^{-1} decreases with decreasing temperature confirming that it is a combination band. Similarly the band at 744 cm^{-1} is a third-order combination band also attributed to vibrational modes of the Se$_8$ molecules. Indeed the entire absorption spectrum of amorphous selenium, with the exception of the band at 135 cm^{-1} and perhaps a weak shoulder at 230 cm^{-1}, can be attributed to vibrational absorption bands of the Se$_8$ molecular species. Note that this is the first definitive evidence for the presence of Se$_8$ molecules in amorphous selenium. A comparison of the Raman spectra of the three forms of selenium yields a similar conclusion. Srb and Vaško[9] have come to a similar conclusion by a more indirect route. They noted similarities between the spectrum of amorphous selenium and amorphous and crystalline (orthorhombic) sulfur.

It would be tempting to try to make quantitative statements about the relative populations of the ring and chain components of the amorphous material. This is at best difficult. Experiments performed by Dr. A. T. Ward of the Xerox Research Laboratories on the Raman spectrum of liquid sulfur above and below the polymerization temperature indicated that the spectrum is dominated by modes associated with the S_8 molecule.[8] Further work is clearly in order if techniques are to be developed which relate the absorption and scattering strength to the fractional population of the constituent species. Preliminary considerations indicate that Raman spectroscopy may be the more promising technique.

IMPURITY ABSORPTION BANDS IN AMORPHOUS SELENIUM

Arsenic

The addition of arsenic to selenium produces several new absorption bands, among which is a combination mode band at 650 cm^{-1}. This band was observed at the same frequency in glassy As_2Se_3 and is attributed to a vibrational mode of that species. The absorption strength of the "As_2Se_3" combination band at 650 cm^{-1} and the strength of the Se_8 combination band at 744 cm^{-1} as a function of arsenic content have been studied (Fig. 8). It was found that the strength of the As_2Se_3 band at 650 cm^{-1} increases linearly with increasing arsenic concentration, whereas the strength of the Se_8 vibration decreases linearly, extrapolating to zero at a 20 per cent concentration of arsenic.

The absorption edges of As_2Se_3 and amorphous selenium occur at very close to the same wavelength. However, the photoconductivity of As_2Se_3 rises at the absorption edge showing no region of strong absorption without photoconductivity as does amorphous selenium. When arsenic is added to selenium the red response of the photoconductor is increased. This increased red response is attributed to the build-up of As_2Se_3. In addition, the non-photoconductive absorption of selenium may be tied up with excitations of the Se_8 molecule. Again further studies are clearly necessary.

Chlorine and Bromine

The absorption spectra of chlorine- and bromine-doped selenium have been studied for concentrations of the order of 50 to 500 ppm. These two impurities are believed to terminate selenium chain ends as evidenced by their similar effects on the viscosity and growth habit of trigonal selenium.[10] For bromine-

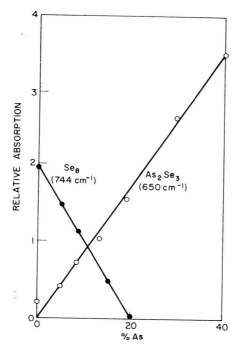

FIG. 8. Relative absorption of combination of Se_8 (744 cm^{-1}) and As_2Se_3 (650 cm^{-1}) combination bands as a function of arsenic content.

doped samples with concentrations to 5000 ppm no additional absorption bands are observed, whereas for chlorine-doped samples additional bands are observed at 340, 550, 590, 720 and 900 cm^{-1} for samples containing 500 ppm chlorine. The absorption strength in these bands increases with decreasing temperature.†

The band at 340 cm^{-1} is attributed to a vibrational mode of a chlorine-terminated chain. The mass of chlorine is lighter than selenium, so localized mode formation is possible. Further note that the mass of bromine is about equal to that of selenium, so that localized mode formation in that system is not expected. Therefore, although chlorine and bromine affect viscosity and growth habit in similar ways, the relationship of their respective atomic masses to that of selenium produces different optical effects.‡

† The band at 340 cm^{-1} is due to chlorine-terminated chains. The sharper bands at 550, 590, 720 and 900 cm^{-1} appear to be related to oxygen since they lie close to those observed in amorphous selenium doped with selenium dioxide and since their intensities decrease when rigorous precautions are taken to exclude oxygen during the preparation of the chlorine-doped specimens.[11]

‡ In the same context, it appears that contamination by oxygen is a less serious problem in the preparation of bromine-doped selenium than it is in the case of chlorine-doped selenium.

ACKNOWLEDGMENT

The author is indebted to R. C. Keezer for the preparation of the amorphous and crystalline selenium samples and for the chlorine- and bromine-doped amorphous selenium samples; to E. J. Felty and M. B. Myers for the preparation of the As–Se alloys; to R. Zallen for helpful discussions and to A. T. Ward for his Raman results, prior to publication, on crystalline and liquid sulfur. All those acknowledged are members of the Xerox Research Laboratories.

REFERENCES

1. LUCOVSKY, G., MOORADIAN, A., TAYLOR, W., WRIGHT, G. B. and KEEZER, R. C., *Solid State Comm.* **5**, 113 (1967).
2. LUCOVSKY, G., KEEZER, R. C. and BURSTEIN, E., *Solid State Comm.* **5**, 439 (1967).
3. CALDWELL, R. S. and FAN, H. Y., *Phys. Rev.* **114**, 664 (1959).
4. HULIN, M., *Proc. Int. Conf. on Lattice Dynamics, J. Phys. Chem. Solids* **21**, (suppl) 135 (1965).
5. BURSTEIN, E., BRODSKY, M. H. and LUCOVSKY, G., *Proc. of the Symposium on Atomic, Molecular and Solid-State Theory*, Sanibel Island, 1967 (to be published).
6. SCOTT, D. W., MCCULLOUGH, J. T. and KRUSE, F. H., *J. Mol. Spectroscopy* **13**, 313 (1964).
7. CHANTRY, G. W., ANDERSON, A. and GEBBIE, H. A., *Spectrochimica Acta* **20**, 1223 (1964).
8. WARD, A. T., unpublished data.
9. SRB, I. and VAŠKO, A., *Czech. J. Phys.* **B13**, 827 (1963).
10. KEEZER, R. C., *The Physics of Selenium and Tellurium*, Pergamon, New York, 1969, p. 103.
11. BURLEY, R. A. and SIEMSEN, K. (Noranda Research Centre), private communication.

DISCUSSION

KOLB: (1) You referred to the heat treatment of amorphous selenium samples and their conversion to trigonal selenium. At what temperature were the samples heat treated? (2) How were the surfaces of the three different forms of selenium prepared for the infrared measurements?

LUCOVSKY: (1) At 80° for 16 hours. (2) (a) Amorphous: cast samples—samples were cast between quartz plates; surfaces were as formed. Evaporated samples—surfaces as grown. (b) α-Monoclinic: as grown. (c) Trigonal: (10$\bar{1}$0) faces were cleaved and etched.

WALLDÉN: Might optical measurements be of any help in studies of macromolecular structure in selenium and the influence of trace elements such as halogens and arsenic, on structure?

LUCOVSKY: Yes, we have indicated how infrared studies of chlorine- and arsenic-doped selenium have given us a model for structural changes in doped material. Measurements near the band edge, onset of electronic transitions, also have aided in an understanding of the structure of arsenic-doped selenium.

HULIN: What were your reasons for attributing to the zero wave vector optical E modes in trigonal selenium, the simple motions shown in your figure, i.e. (1) motion $\perp$ c-axis (two degenerate levels); (2) motion $\parallel$ c-axis (two degenerate levels)?

LUCOVSKY: As you pointed out, the actual motions for the E modes are linear combinations of the simple motions shown. The large separation of these modes in frequency space, i.e. one at 230 cm^{-1} and one at 144 cm^{-1}, is indicative of a small coupling via electrostatic forces. The argument is favored by the near degeneracy of the A_1 and high-frequency type E mode.

VAŠKO: We found a weak infrared absorption band in amorphous selenium at 77 cm^{-1}, which we attributed to a Raman-active ring mode. Infrared activity for this mode is derived from a breakdown of selection rules due to intermolecular interactions in the solid. Do you agree with this interpretation?

LUCOVSKY: I agree with your interpretation and remark that this weak band is observable only in very thick samples. We did not observe it in the thin (4×10^{-2} cm) samples which we studied in the far infrared.

LANYON: Regarding the decrease in the intensity of the Se_8 absorption band with the increase in the intensity of the As_2Se_3 band. (1) Can one assume from this that the first effect of the incorporation of arsenic is to break up the Se_8 molecules prior to their incorporation into the selenium chains? (2) Would this effect enable one to estimate the relative population of selenium atoms in rings and chains in view of the difficulty involved in using the relative intensities of the bands themselves?

LUCOVSKY: (1) Yes, this appears to be the case. The Se_8 ring is less stable than the chain and appears to interact strongly with the arsenic atoms. (2) If we assume an interaction of the form

$$16As + 3Se_8 \rightarrow 8As_2Se_3$$

then 20 per cent arsenic addition corresponds to 30 per cent of the selenium atoms in Se_8 rings. This estimate is in accord with ring-chain equilibrium theories for the ring-chain ratio at the melting point of selenium. However, it is an estimate.

BECKER: (1) Can you distinguish between chains and very large rings in amorphous selenium via infrared spectroscopy? (2) Can you give semiquantitative estimates of the number of atoms in rings (Se_8) and chains?

LUCOVSKY: (1) No, the infrared activity of vibrational modes in a large ring structure ($n \sim 100$) would be essentially the same as that of a chain. It would be dominated by the mechanical motions of a three-atom repeat group. (2) Only with extreme caution. Ring modes dominate both the infrared and Raman spectra. However, we have no clear-cut way of correlating absorption band strengths with ring and chain atomic populations. Measurements of the Raman spectrum of liquid sulfur are being performed by A. T. Ward of the Xerox Research Laboratories and studies of this sort should help us give a quantitative answer to this question.

THE RAMAN SPECTRUM OF TRIGONAL, α-MONOCLINIC AND AMORPHOUS SELENIUM†

A. MOORADIAN and G. B. WRIGHT

Lincoln Laboratory, Massachusetts Institute of Technology, Lexington, Massachusetts

INTRODUCTION

The vibrational modes of trigonal, α-monoclinic, and amorphous selenium have been studied in the past by several workers.[1-4] A preliminary account of the results of the present Raman investigation has appeared in a recent paper[4] in which identification of the vibrational modes of the three forms of selenium was made by combining results from Raman and infrared measurements. In the present study the Raman spectrum of single crystal[5] trigonal and α-monoclinic, and amorphous selenium measured at room temperature and liquid helium temperature has been investigated.

EXPERIMENTAL

A 1.06 micron Yttrium Aluminum Garnet: Neodymium laser operating at room temperature with an output of 3 watts CW was used as the source. The experimental arrangement is shown in Fig. 1. The scattered light was collected at 90 degrees from the direction of the laser beam and analyzed by a tandem grating spectrometer with gratings blazed at one micron. For the weakest signals, photon counting was employed using a cooled S-1 phototube.

RESULTS

Trigonal Selenium

The crystal structure of trigonal selenium consists of parallel spiral chains at the corners and center of a hexagon. Every third atom in a chain completes one revolution of the spiral so that the projection of the atoms on a plane perpendicular to the chain axes consists of equilateral triangles. The c-axis is along the chains. There are three atoms per unit cell.

Trigonal selenium has D_3 point group symmetry and group theory predicts a non-degenerate A_1 Raman active mode, two doubly degenerate E Raman and infrared active modes, and a non-degenerate A_2 infrared active mode.

† This work was sponsored by the U.S. Air Force.

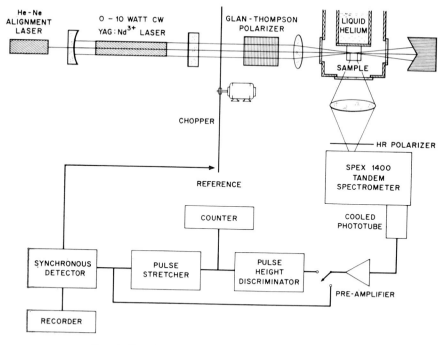

FIG. 1. Experimental setup for Raman spectroscopy.

A recorder trace of the first-order Raman spectrum of trigonal selenium taken at room temperature is shown in Fig. 2. The sample was polycrystalline and the laser radiation unpolarized. The system response drops toward longer wavelengths which accounts for the anti-Stokes components having larger amplitudes than the Stokes components. Figure 3 shows recorder traces of the first-order modes taken at helium temperature. The lines narrow from a half

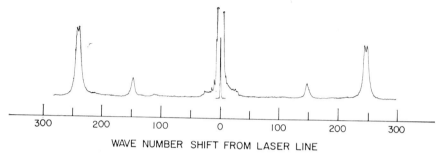

FIG. 2. Recorder trace of the first-order Raman spectrum of polycrystalline trigonal selenium measured at room temperature. (Note: the wave number scale on this and all other figures is non-linear.) The laser was unpolarized.

width of 5 cm^{-1} at room temperature down to 2 cm^{-1} at liquid helium temperature which corresponds to an increase of phonon lifetime from 1.2×10^{-11} sec to 3×10^{-11} sec at helium temperature.

The identification of the modes by polarized light is shown in Fig. 4, where a polarized laser and a single crystal sample were used. Since trigonal selenium is strongly optically active along the c-axis, both incident and scattered light were at right angles to the c-axis to keep rotation effects to a minimum. The Raman scattering tensors for D_3 point group symmetry have been listed by Loudon[6] and permitted the calculation of the polarization properties of the

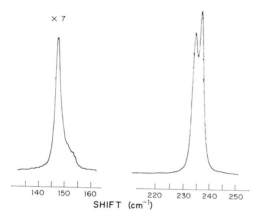

FIG. 3. Recorder spectra of the first-order Stokes modes in polycrystalline trigonal selenium taken at liquid helium temperature. Resolution is about 0.3 cm^{-1}.

scattered light. In Fig. 4 for the laser polarized perpendicular to the plane of scattering, only the two E modes are present, while for the laser polarized parallel to the plane of scattering, only the A_1 mode is prominent.

The second-order Raman spectrum of trigonal selenium is shown in Fig. 5 at room temperature and liquid helium temperature. The line at 102 cm^{-1} narrowed considerably at helium temperature. This peak occurs at the same frequency as the infrared active A_2 mode identified in the infrared absorption.[4] Although the A_2 mode is not Raman allowed from group theoretical selection rules, it is apparently turned on by the presence of some crystal imperfection.

Much of the remaining structure can be attributed to the simultaneous emission of two phonons of equal and opposite wave vector. If it is assumed that the probability for the two phonon Raman process is independent of wave vector, then the structure in the second order spectrum reflects the peaks in the density of states of the phonon dispersion curves.

Hulin[7] has published a description of the lattice dynamics of tellurium based on a Born–von Karman force model, and recently Geick et al.[8] have

272 A. MOORADIAN AND G. B. WRIGHT

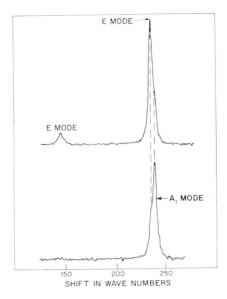

FIG. 4. Raman spectrum of single crystal trigonal selenium at room temperature. The laser was polarized perpendicular (upper trace) and parallel (lower trace) to the plane formed by the incident and scattered light directions.

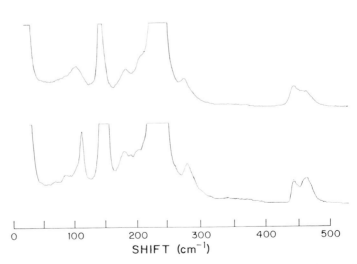

FIG. 5. Recorder traces of the second-order Raman spectra of polycrystalline trigonal selenium at room temperature and liquid helium temperature,

extended the model to calculations of the dynamics of selenium. The latter model is an empirical fit to the infrared, Raman, and elastic constant data for selenium, and achieves reasonable agreement with a seven parameter model. Table 1 lists the observed Raman frequencies and the associated mode identifications.

TABLE 1. *Observed Raman Lines and Symmetry Assignment in Trigonal Selenium*

Fundamental Modes

Observed frequencies (cm^{-1})		Symmetry assignment
Room temp.	Liq. helium temp.	
102	112	A_2
143	147	E
233	232	E
237	235	A_1

Observed Second-order Peaks (cm^{-1})

Room temp.	Liquid helium temp.	
75	85	220
145	100	280
183	153	341
206	180	366
273	192	394
345 (broad band)	205	442
438	212	460
455		

α-Monoclinic Selenium

This form of selenium has a unit cell containing four Se$_8$ molecules held together by van der Waal's forces. An approach to the analysis of such a complicated structure may be made by comparing the spectrum to that of the Se$_8$ ring molecules studied by Scott and co-workers.[9] The results of such an analysis are given in Table 2. The data for α-monoclinic selenium are shown in Fig. 6 at room temperature and at helium temperature. There is a radical sharpening of the spectrum with decrease in temperature. While there is reasonable confidence in the assignments attributed to ring molecule vibrations, the nature of the lattice vibration modes requires a much more elaborate analysis. Identification of the ring vibrations is very helpful for the analysis of the amorphous selenium spectra, however,

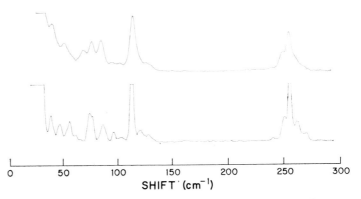

FIG. 6. Recorder traces of the Raman spectra of α-monoclinic selenium at room temperature and liquid helium temperature.

TABLE 2. *Observed Raman Lines and Symmetry Assignment in α-Monoclinic Selenium at Liquid Helium Temperature*

Observed frequency shift (cm^{-1})	Se_8 mode assignment
38	—
47	E_2
55	—
60	—
74	—
76	—
86	E_2
96	E_1
103	—
113	A_1
120	B_2
128	E_3
162	—
176	—
240	E_3
251	A_1
256	E_2
264	E_1
271	—

Amorphous Selenium

This form of selenium is composed of a mixture of Se_8 ring molecules and long chains similar to those found in the trigonal form. Evidence for the existence of both the Se_8 ring and trigonal chain modes is presented in Fig. 7

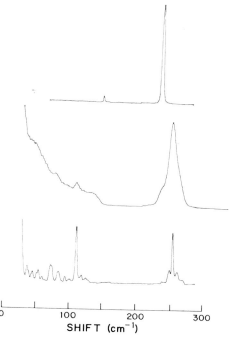

FIG. 7. Recorder traces of the Raman spectra of trigonal (top) α-monoclinic (bottom) and amorphous selenium (middle) at liquid helium temperature.

where the first-order Raman spectra of amorphous, α-monoclinic, and trigonal selenium at liquid helium temperature are compared. While the trigonal and α-monoclinic spectra sharpen at the lower temperatures, there was no appreciable line narrowing in the amorphous spectrum in going to liquid helium temperature. This appears consistent with the disordered nature of amorphous selenium.

ACKNOWLEDGMENT

The authors thank D. Wells for assistance with the measurements, R. Geick, U. Schröder and J. Stuke for providing a preprint of their paper prior to its publication, and R. C. Keezer for providing the single crystal of trigonal selenium.

REFERENCES

1. CALDWELL, R. S. and FAN, H. Y., *Phys. Rev.* **114**, 664 (1959).
2. CHANTRY, G. W., ANDERSON, A. and GEBBIE, H. A., *Spectrochimica Acta* **20**, 1223 (1964).
3. SRB, I. and VAŠKO, A., *Czech. J. Phys.* **B13**, 827 (1963).
4. LUCOVSKY, G., MOORADIAN, A., TAYLOR, W., WRIGHT, G. B. and KEEZER, R. C., *Solid State Comm.* **5**, 113–117 (1967).

5. KEEZER, R. C., WOOD, C. and MOODY, J. W., *Proc. Int. Conf. on Crystal Growth*, Pergamon, New York, 1967.
6. LOUDON, R., *Advan. Phys.* **13**, 423 (1964).
7. HULIN, M., *Ann. Phys. (N.Y.)* **8**, 647 (1963).
8. GEICK, R., SCHRÖDER, U. and STUKE, J., *Phys. Status Solidi* **24**, 99 (1967).
9. SCOTT, D. W., MCCULLOUGH, J. T. and KRUSE, F. H., *J. Mol. Spectroscopy* **13**, 313 961)('4

DISCUSSION

LACOURSE: Do you feel that your findings and those of Lucovsky on infrared spectra would rule out the existence of Se_6 rings in amorphous selenium?

LUCOVSKY: (a) It would be surprising if both the Raman and infrared frequencies of six- and eight-membered rings were identical. One would expect higher frequencies in six-membered rings.

(b) There is no evidence for a solid form of selenium that is characterized by a six-membered ring. Radial distribution studies on amorphous selenium have been interpreted and have yielded a possible cross-linked structure based on six-membered rings. However, this structural model relies heavily on higher order next nearest neighbor distances 5, 6, 7, 8, etc., and is somewhat questionable.

GEICK: Were you able to identify any Brillouin components in your Raman scattering set up?

MOORADIAN: The experimental setup did not have sufficient resolution to observe Brillouin components which would be expected to have a shift of only about one wave number.

WOOD: Could you distinguish between α- and β-monoclinic selenium spectra?

MOORADIAN: A comparison of the Raman spectra of α- and β-monoclinic selenium could not be made because we had no β-monoclinic selenium samples.

THE PHONON SPECTRA OF TRIGONAL SELENIUM AND TELLURIUM

R. GEICK

Physikalisches Institut der Universität Freiburg, Freiburg, Germany

and

U. SCHRÖDER

Institut für Theoretische Physik der Universität, Frankfurt, Germany

INTRODUCTION

Considerable interest has been shown recently in the lattice vibrations and the lattice dynamics of the homopolar semiconductors selenium and tellurium. Phonon frequencies have been determined by means of infrared and Raman spectroscopy and by means of inelastic scattering of neutrons. Also the values of the elastic constants have been measured.

Spectroscopic investigations, namely reflection and transmission measurements, yield only the frequencies of the infrared- and Raman-active phonons for nearly zero wave vector because of wave vector conservation. For trigonal selenium and tellurium, five of the six optical modes for $q \approx 0$ are infrared-active (two pairs of doubly degenerate modes for $E \perp c$ and one non-degenerate mode for $E \| c$), and the sixth mode is Raman-active only. The dipole moment associated with the infrared-active lattice vibrations originates in this case from the deformation of the electron shells and not from the displacement of charged ions. For selenium, all four distinct frequencies for $q \approx 0$ were obtained from infrared data and from the Raman effect,[1-4] while for tellurium only infrared data are available.[1, 3, 5, 6]

In addition to the absorption lines due to the infrared-active phonons, several minor broad bands are found in the absorption spectra which are caused by higher-order processes, i.e. two- and more-phonon absorption. In these processes, phonons throughout the Brillouin zone can take part. The two-phonon bands result from a sum of all processes possible at the frequency of the incident light, and thus they do not reveal specific information about a particular phonon.

The inelastic coherent scattering of thermal neutrons offers a possibility to determine the phonon dispersion curves in dependence on the wave vector. But such data are not yet available for single crystal selenium and tellurium due to the lack of sufficiently large crystals. However, for tellurium investigations of this kind are in progress.[7] At present, only inelastic scattering by polycrystalline samples has been studied.[8] The results represent an average at a given energy and thus will reveal the essential features of the one-phonon density of states. Particularly the peaks in the time-of-flight spectrum may be assigned to the phonons at points in reciprocal space with a high density of states (critical points). Five such peaks have been found as well for selenium as for tellurium. Three peaks are easily identified as originating from optical branches while the other two arise from acoustical branches.

A third possibility to determine phonon frequencies experimentally is given by the measurement of sound velocities and elastic constants. These data represent detailed information about the phonon dispersion curves for long acoustic waves near the center of the Brillouin zone. For tellurium, a complete set of elastic constants has been published several years ago.[9] For selenium, in addition to the determination of c_{33} and c_{44},[10] the velocities of sound waves travelling perpendicular to the c-axis have been studied very recently.[11]

The conclusion from these considerations is that the various available experimental data either reveal information about phonons near the center of the Brillouin zone or represent only integrated information about a large number of phonons at the edge of the zone. Therefore it is difficult to compare these data with each other and to combine them to a complete picture of the lattice dynamics of selenium or tellurium. In order to overcome this difficulty it seemed advisable to undertake a calculation of the phonon frequencies by means of a simple force constant model. Some of the experimental data will be used to adjust the parameters in the model calculation while the other data may be compared with the corresponding calculated ones.

FORCE CONSTANT MODEL AND CALCULATION OF PHONON SPECTRA

The model chosen for the calculation of the lattice frequencies is an extension of that employed by Hulin[12] who used central forces between nearest and next nearest neighbors and angular forces including third nearest neighbors. The configurations of first, second and third neighbors in the trigonal structure of selenium or tellurium are shown in Fig. 1. For the present calculation, symmetry coordinates were introduced which are invariant against a threefold rotation (cf. Fig. 1). Along the chains the most general force constants consistent with symmetry and invariance requirements were used for the forces between nearest and third nearest neighbors. For the forces between next nearest neighbors which belong to adjacent chains, a central force model was employed.

After reducing the number of free parameters by the use of symmetry and the usual invariance requirements,[13] seven parameters remain to be determined from experimental data in this model.†

When this work was started only the values of six quantities were available for this purpose in the case of selenium. Therefore, the number of free parameters in the model was finally reduced to six assuming the constants for the z-component of the forces arising from r- or ϕ-components of the displacements, respectively, not to be independent. For selenium, the six values to fit the

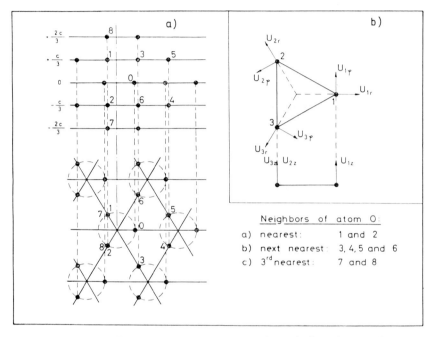

FIG. 1. Position of nearest, next nearest and third neighbors in the trigonal structure of selenium and tellurium (a), and the symmetry coordinates used for the model calculation (b).

parameters were the four frequencies of the optical phonons for $q \approx 0$ and the values of the two elastic constants c_{33} and c_{44}. For tellurium, the values of the same quantities were used with one exception: instead of the frequency of the mode which is Raman active, only the value of another elastic constant, namely c_{13}, had to be used. While in the case of tellurium the evaluation of the force constant parameters from the experimental data was straightforward, for selenium this relatively simple model proved to be not quite consistent with the actual data. Therefore, such values for the parameters were chosen that an optimum fit of the calculated values to the experimental data is achieved. In this

† For details see ref. 4.

procedure, differences of about 10 per cent between calculated and experimental values had to be tolerated. The input data used for the model calculation are compiled in Table 1.

TABLE 1. *Input Data for the Calculation of Phonon Spectra*

	Selenium	Tellurium
Frequencies of optical phonons for $q \approx 0$ (10^{13} sec^{-1})		
ω_4 (A_2)	1.95[b]	1.62[c]
$\omega_{5,\,6}$ (E)	2.63[b]	1.77[c]
$\omega_{7,\,8}$ (E)	4.39[b]	2.67[e]
ω_9 (A_1)	4.46[a]	
Elastic stiffness constants (10^{11} dynes/cm^2)		
c_{33}	8.02[d]	7.22[e]
c_{44}	1.83[d]	3.14[e]
c_{13}	—	2.50[e]

[a] Taken from ref. 2. [b] Taken from ref. 4. [c] Taken from ref. 5. [d] Taken from ref. 10. [e] Taken from ref. 9.

The results of the model calculation, the phonon dispersion curves for certain directions of the Brillouin zone, are shown in Fig. 2 for selenium, and in Fig. 3 for tellurium. The dispersion of the acoustical branches (A) and the lower set of optical branches (O) is very similar to the dispersion of acoustical and optical phonons in other lattices while the upper optical branches (O') exhibit a dispersion more like that of a molecular vibration. For a number of lattice modes the chains behave as rigid bodies, and the frequencies of these modes depend only on the forces between different chains. The corresponding force constants turn out to be smaller by a factor 5 to 10 than the forces reacting on a stretching of the covalent bond for selenium as well as for tellurium. At $B = (4\pi/3a)$, O, O for the acoustical branch with the lowest energy, the frequency vanishes (cf. Figs. 2 and 3). In our model with central forces between adjacent chains, even these forces compensate each other for the phonons under discussion thus showing that a more complex model has to be employed. This weak point of the model, and also of the model of Hulin,[12] can be overcome and the lattice stabilized by adding forces between fourth nearest neighbors.

The phonon spectra from the model calculation now will be compared to other data known about the lattice dynamics of selenium and tellurium which have not been used to determine the force constants of the model. For selenium, the calculated frequencies agree quite well with the peaks obtained from inelastic scattering of neutrons by polycrystalline samples,[8] particularly the

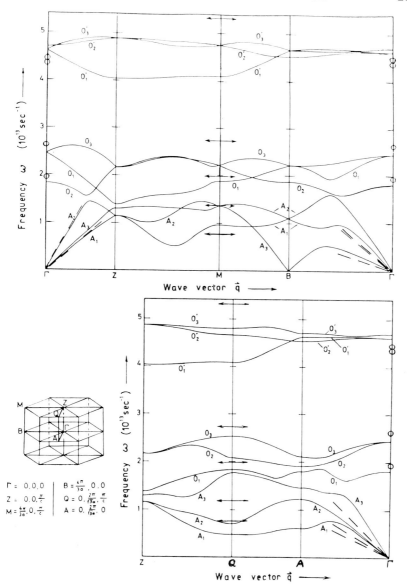

FIG. 2. Phonon dispersion curves of trigonal selenium versus wave vector for certain directions in the Brillouin zone indicated in the figure. The phonon branches have been numbered A_1, A_2 and A_3 (acoustical branches); O_1, O_2 and O_3 (first set of optical branches); and O'_1, O'_2 and O'_3 (second set of optical branches). The experimental data are given in the figure as follows: phonon frequencies at $q \approx 0$ (O), elastic constants (– – –) and peaks from neutron spectrometry data (←→).

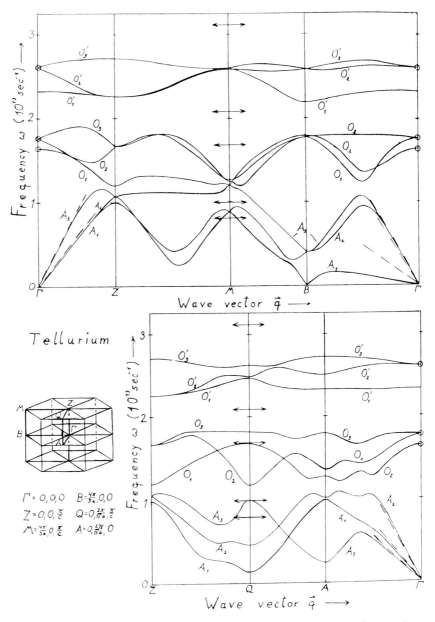

FIG. 3. Phonon dispersion curves of tellurium versus wave vector for certain directions in the Brillouin zone indicated in the figure. The phonon branches have been numbered A_1, A_2 and A_3 (acoustical branches); O_1, O_2 and O_3 (first set of optical branches); and O_1', O_2' and O_3' (second set of optical branches). The experimental data are given in the figure as follows: phonon frequencies at $q \approx 0$ (O), elastic constants (- - -) and peaks from neutron spectrometry data ($\leftrightarrow$).

neutron peaks assigned to the A- and O-branches fit well to these branches at the edge of the Brillouin zone (cf. Fig. 2). Only the energies of the O'-branches are by about 10 per cent lower than the frequency of the corresponding peak from the neutron data. On the other hand, there are considerable discrepancies between the calculated values of the elastic constants c_{11}, c_{12} and c_{14}[11] and the experimental data (cf. Fig. 2).

In contrast to that, for tellurium the agreement between the sound velocities in the present model and the experimental values is quite good except for the lowest acoustic branch between Γ and B because of the instability already discussed. But here the calculated phonon frequencies at the edge of the Brillouin zone turn out to be lower than the corresponding neutron peaks. Also the calculated frequency of the Raman-active mode at Γ seems to be somewhat low although no experimental value is available for comparison in this case. Summarizing these considerations, the model calculation for selenium reproduces within 10 per cent the phonon frequencies of the optical branches while the model calculation for tellurium shows a better agreement with the elastic data. Now, one force constant was determined from the Raman frequency in the case of selenium and from c_{13} in the case of tellurium. Therefore it seems that this relatively simple model cannot describe the dynamics of optical and acoustical phonons simultaneously in an adequate way. This weak point most probably arises from the fact that long-range Coulomb forces have been neglected. The addition of these forces (e.g. by means of a shell model) will alter the frequencies of the optical branches considerably but will have little influence on the elastic properties.

TWO-PHONON SPECTRA AND COMBINED DENSITY OF STATES

Other data to be discussed are the two-phonon peaks in the infrared and Raman spectra. As is well known, the absorption coefficient for two-phonon absorption depends on the combined density of states of the two branches and on coupling parameters which couple the phonons to the electromagnetic field either by anharmonic decay of an infrared-active phonon or by nonlinear dipole moment.[13, 14] Both mechanisms yield the same selection rules. Now, an appropriate way of discussing the two-phonon spectra is comparing these to the combined density of states. In this way coupling parameters and selection rules are neglected for the present. Using the force constant model, the phonon frequencies have been calculated for 12,288 points distributed uniformly over the Brillouin zone, and the combined density of states for two-phonon summation processes has been evaluated for selenium (cf. Fig. 4) and for tellurium (cf. Fig. 5). In the frequency ranges under consideration summation processes will be more important than difference processes.

For selenium (cf. Fig. 4), three broad bands with a high combined density arise for reasons of the relatively large gap between O and O' branches in the phonon spectrum. The positions of these bands agree quite well with the two-phonon peaks from infrared and Raman data.[2, 4] Even the small gap between the peaks at about 320 cm^{-1} corresponds well to a relative minimum in the combined density. In this way, all two-phonon peaks can be explained in

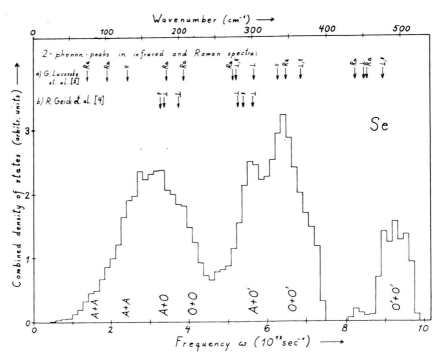

FIG. 4. Combined density of states for two-phonon summation processes in trigonal selenium, and for comparison, two-phonon peaks from infrared and Raman spectra indicated by arrows.

an unconstrained way including those peaks which are not easily explained in terms of the zone center values of the optical frequencies. Besides the number of two-phonon combinations possible at a given frequency, selection rules required by symmetry have to be considered in the analysis of two-phonon spectra. Now, such selection rules are easily derived for symmetry points Γ, Z, M and B in reciprocal space.[12, 15] And for a first approximation, conclusions will be drawn for the whole spectrum from selection rules valid at these special points. For example, the overtones of branches nondegenerate at M or B are Raman-active only ($M_1 \times M_1 = \Gamma_1$, $M_2 \times M_2 = \Gamma_1$). This would explain why the peaks at 75 cm^{-1} ($A+A$) and at 438 cm^{-1} ($O'+O'$) have been found by Lucovsky et al.[2] in the Raman spectrum only and not in the infrared spectra. However, most of the two-phonon peaks have been found in Raman spectra as well as in infrared spectra because combinations of branches with all possible symmetry characters contribute to these peaks. It can be shown for M and B that the combination branches infrared-active for $E\|c$ exhibit less dispersion than those active for $E\bot c$. Therefore, the two-phonon

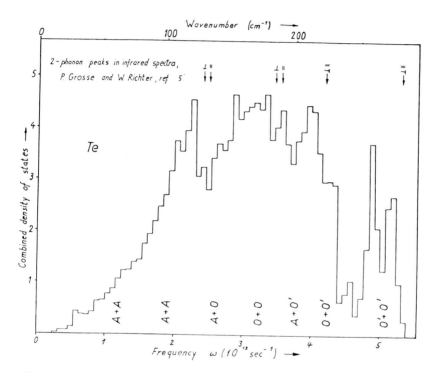

FIG. 5. Combined density of states for two-phonon summation processes in tellurium, and for comparison, two-phonon peaks from infrared absorption spectra indicated by arrows.

bands for $E\|c$ are expected to be narrower than the bands for $E\bot c$. And in fact, the two-phonon peaks for selenium at about 180 cm^{-1} and 280 cm^{-1} are narrower for $E\|c$ than for $E\bot c$.[4]

In the case of tellurium, the structure of three broad bands in the combined density of states is not quite so pronounced as for selenium. In comparison to the two-phonon peaks from infrared spectra,[5] the peaks in the combined density appear at somewhat lower frequencies. This indicates once more that the calculated phonon spectrum of tellurium is shifted to lower frequencies in comparison with experimental data as already discussed.

ACKNOWLEDGMENT

The authors are grateful to Prof. Dr. H. Bilz for helpful and valuable discussions and gratefully acknowledge the encouragement and other support of Prof. Dr. L. Genzel and Prof. Dr. J. Stuke.

REFERENCES

1. CALDWELL, R. S. and FAN, H. Y., *Phys. Rev.* **114**, 664 (1959).
2. LUCOVSKY, G., MOORADIAN, A., TAYLOR, W., WRIGHT, G. B. and KEEZER, R. C., *Solid State Comm.* **5**, 113 (1967).
3. LUCOVSKY, G., KEEZER, R. C. and BURSTEIN, E., *Solid State Comm.* **5**, 439 (1967).
4. GEICK, R., SCHRÖDER, U. and STUKE, J., *Phys. Status Solidi* **24**, 99 (1967).
5. GROSSE, P., LUTZ, M. and RICHTER, W., *Solid State Comm.* **5**, 99 (1967).
6. GROSSE, P. and LUTZ, M., *Verhandl. DPG* (6) **2**, 30 (1967).
7. AXMANN, A., GISSLER, W. and SPRINGER, T., private communication.
8. AXMANN, A. and GISSLER, W., *Phys. Status Solidi* **19**, 721 (1967).
9. MALGRANGE, J. L., QUENTIN, G. and THUILLIER, J. M., *Phys. Status Solidi* **4**, 139 (1967).
10. VEDAM, K., MILLER, D. L. and ROY, R., *J. Appl. Phys.* **37**, 3432 (1966).
11. MORT, J., *J. Appl. Phys.* **38**, 3414 (1967).
12. HULIN, M., *Ann. Phys. (Paris)* **8**, 647 (1963).
13. BORN, M. and HUANG, K., *Dynamical Theory of Crystal Lattice*, Clarendon Press, Oxford, 1954, Section V.
14. SZIGETI, B., *Proc. Roy. Soc.* A **258**, 377 (1960).
15. NUSSBAUM, A., *Proc. IRE* **50**, 1762 (1962).

DISCUSSION

LUCOVSKY: The relative importance of anisotropy on electrostatic forces can be estimated by comparing the A_2-E (Stronger) splitting in selenium and tellurium. In selenium the splitting of the respective TO and LO frequencies of the two modes is much larger than the $LO-TO$ splitting of either mode, whereas in tellurium the corresponding $LO_E-LO_{A_2}$, $TO_E-TO_{A_2}$ splitting is comparable to the $LO_{A_2}-TO_{A_2}$, LO_E-TO_E splitting of either mode. Therefore electrostatic forces are more important than anisotropy in tellurium as compared to selenium. The long-range Coulomb forces in tellurium shift the E mode down and pull the A_2 mode up (in frequency) as evidenced by comparing the measured frequencies for tellurium, 90 (A_2)–92 (E), with the calculated values (74) A_2 and (122) E from Hulin's calculation which neglects electrostatic forces. Perhaps this is the origin of the better agreement between your calculated density of states and the experimental second order bands of selenium as compared to tellurium, i.e. you have neglected electrostatic forces which are of less importance in selenium as compared to tellurium.

GEICK: Adding Coulomb forces to the model may remove the discrepancies found with the simple force constant model in comparison to experimental data. However, it may turn out then that for the short-range forces a more complex model has to be employed in order to take account of the anisotropy in the trigonal structure.

BECKER: The discussion has pointed out the need for a shell model calculation. I merely wish to point out that Dr. Inan Chen of Xerox is currently carrying out such a calculation for selenium.

HULIN: Did you check that the symmetry requirements and the conditions for absence of external stresses were enough to exclude first-order force constants?

GEICK: The model was derived from the equations of motion rather than from a potential. The relatively large number of force constants was reduced by symmetry requirements and the general invariance requirements (absence of external stresses, no forces arising from displacement and rotation of the lattice as a rigid body). Then no evidence was found for the existence of first-order force constants.

HULIN: Would using the new experimental values for the elastic coefficients (Mort et al.) remove the discrepancies between experimental and theoretical values for acoustical phonon energies?

GEICK: Certainly not because there is not much difference in the values of the elastic constants c_{33} and c_{44} between the results of Vedam et al. and of Mort, and there is no disposable parameter left in our model to fit the calculated sound velocities in (100) direction, for example, to the experimental data.

PHOTOLUMINESCENCE OF SELENIUM AND SELENIUM–TELLURIUM MIXED CRYSTALS

H. J. QUEISSER

Physikalisches Institut der Johann Wolfgang Goethe-Universität, Frankfurt, Germany

LUMINESCENCE studies present a powerful tool to obtain detailed information about the electronic states within the forbidden energy gap of semiconductor crystals. Thus far very little is known about electronic states of impurities and defects in selenium single crystals. This paper describes some results of photoluminescence measurements on selenium crystals and some selenium–tellurium mixed crystals. Luminescence spectra have been observed between 2°K and about 50°K. The interpretation of the spectra is, however, difficult since at present little correlation can be made to other measurements. The work to be presented was initiated at the Bell Telephone Laboratories[1] and is now being continued and extended at Frankfurt University.[2]

The experimental setup utilizes a helium–neon laser as a light source for exciting hole-electron pairs in specimens. The laser provides 1.96 eV photons with high spectral purity. In particular, there is a very low background in the spectral region of the luminescence. This is vital to the success of the experiments since quantum efficiencies are very low. The laser is focused on the crystal which is held in a cryostat; in many cases it is directly immersed in the refrigerant liquid. Care must be taken to avoid damage to the crystals by the incident exciting radiation. The luminescence can be markedly and irreversibly reduced by power levels as low as 10 mW. The luminescence radiation from the excited front surface of the crystals is focused into a grating monochromator. A cooled photomultiplier with S-1 surface is used in combination with a lock-in amplifier.

A large variety of trigonal crystals have been investigated by this method. Melt-grown selenium–tellurium mixed crystals were investigated in the selenium-rich end of the phase diagram with up to 40 mole per cent tellurium. At 20°K all these crystals show merely broad luminescence bands between 1.2 and 1.8 eV photon energy. The luminescence peaks can be correlated with peaks of the photoconductivity.[3] There is, however, a definite inhomogeneity in these crystals. It was necessary to scan with the exciting laser spot until a sufficiently luminescent region was found. A similar behavior was also found in pure selenium crystals grown from the melt. These crystals show broad bands near 1.65 and 1.8 eV, which have tentatively been assigned to lattice defects.[1] Some crystals indicate a low-intensity spectrum of discrete luminescence lines. Such discrete spectra are particularly prominent in the more perfect vapor-grown needles and platelets. In this paper the discussion will be restricted to the trigonal vapor-grown crystals, since they seem to offer the greatest prospect of furnishing detailed and definitive information.

Figure 1 shows a spectrum at 20°K, which has been reported previously.[1] A triplet A, B, C is seen with some further structure indicated but not resolved. Replicas A', B', C' are observed, each being found at energies 29 meV lower

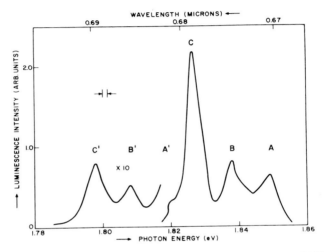

FIG. 1. Luminescence spectrum of trigonal selenium crystals at 20°K.

than the more intense lines of the triplet. The interpretation of the replica as being caused by simultaneous emission of one Raman-active high-energy optical phonon is now in good agreement with the results of measurements in the infrared.[4, 5] Spectra such as shown in Fig. 1 can be observed up to 50°K at which temperature the signals disappear in the noise.

At 4.2°K the spectrum of Fig. 2 is observed. Six lines are now resolved with phonon replicas. Each of the three lines observed at 20°K appears to be accompanied by a new line at approximately 4 meV higher energy. The quantum efficiency is not markedly improved over that at 20°K. The relative intensities of the lines of such a pair can vary greatly from crystal to crystal and within a given sample. At 2°K the spectrum looks surprisingly similar to that obtained at 20°K: only three lines are observed again, with the companion lines merely causing bumps in the high-energy shoulders.

Polarization effects were investigated. With $E \| c$ exciting light the sharpest spectra are found. $E \perp c$ excitation adds a broad response as a background to the lines. It is not clear yet whether this is an effect of the variation of absorption coefficient and consequent surface recombination[1] or whether it is a result of different mechanisms related to band structure and selection rules. The polarization of the emitted luminescence was also checked,[2] although this is difficult because of strong depolarization. It now appears that all lines are polarized $E \perp c$.

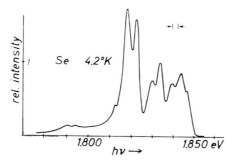

FIG. 2. Spectrum of trigonal selenium at 4.2°K.

The intensity of the luminescence depends in most cases linearly upon the excitation intensity.

Interpretation of the spectra must be done with great caution at the present time. It is not unreasonable to make the assumption of localized levels as caused by impurities or more likely lattice defects to which excitons might be bound. The observed polarization would be in agreement with wave functions made up of the bands responsible for the allowed optical transitions of the $E \perp c$ band gap. The unusual temperature behavior could be the result of a shift of the Fermi level, assumed to be located near two levels close to the valence band edge. It is hoped that further results will clarify these points, and that correlations can be made to other optical data and to electrical data.

REFERENCES

1. QUEISSER, H. J. and STUKE, J., *Solid State Comm.* **5**, 75 (1967).
2. ZETSCHE, H. and FISCHER, R., to be published in *J. Phys. Chem. Solids*.
3. STUKE, J., private communication.
4. GEICK, R., *The Physics of Selenium and Tellurium*, Pergamon, New York, 1969, p. 277.
5. LUCOVSKY, G., *The Physics of Selenium and Tellurium*, Pergamon, New York, 1969 p. 255.

DISCUSSION

LAUDISE: Radiation damage at low laser powers much similar to that which you report has been observed (see, for example, Ashkin, A., Boyd, G. D., Dziedzic, J. M., Smith, R. G., Ballman, A. A., Levinstein, H. J. and Nassau, K., *Appl. Phys. Letters* **9**, 72 (1966)) in a variety of materials including lithium niobate and lithium tantalate.

RICCIUS: Is there any explanation for the observed radiation damage at these low power levels of the He–Ne laser? Usually one finds only radiation damages by using Q-switched lasers.

QUEISSER: Dr. Laudise in his comment gives references concerning similar effects in niobates and other crystals. The damage occurs most probably through local heating. Sometimes one can detect spots on the surface which suggest an effect similar to a local melting of the crystal. We hope to correlate the appearance of luminescence bands with such damage.

KOLB: Since you state that it is possible to see a spot on the surface of a selenium crystal which has suffered laser damage, would it not be possible to lap and polish the surface to remove the spot and examine the sample in crossed polarized light for index of refraction changes?

QUEISSER: This is certainly a good suggestion. We have, however, not attempted to do such measurements. Our main interest thus far was to obtain luminescence spectra with as great a resolution as possible rather than study the laser-induced damage.

LIGHT BEAM MODULATION IN TRIGONAL SELENIUM

J. E. ADAMS and W. HAAS

Research and Engineering Center, Xerox Corporation, Webster, N.Y.

THE magnitude of the linear electro-optic effect in trigonal selenium has been determined. Intensity modulation of 5 per cent was produced using a 1.15 micron helium–neon laser and fields of the order of 10^4 volts/cm.

In isotropic materials at optical frequencies the electric displacement vector **D** is related to an applied field **E** through the dielectric constant K which when combined with Maxwell's equations results in a single propagation velocity in the crystal for a light beam. In anisotropic materials, however, **D** is related to **E** through the permittivity tensor and in this case Maxwell's equations show that for a given propagation direction two waves of mutually perpendicular polarization and different propagation velocity are allowed. The waves are called the ordinary and extraordinary rays or simply O and E rays. This phenomenon of birefringence occurs naturally in crystals of symmetry lower than cubic.

Because the permittivity tensor itself is field dependent the amount of birefringence a crystal exhibits depends on the electric field. The general dependence can be expressed as a power series in the electric field where the quadratic term corresponds to the Kerr effect and the linear term corresponds to the Pockels effect or the linear electro-optic effect. A necessary condition for the Pockels effect is that the crystal possesses no center of symmetry. Because of the tensorial property of the permittivity, the effect of an electric field on the optical properties of a crystal is described in general by an eighteen-component contracted tensor.

However, the symmetry of the crystal limits the number of independent components of this tensor and in the case of class 32 materials, the number of independent components is reduced to 2, and for the special case of light traveling along the c-axis or optic axis and an applied **E** field perpendicular to the c-axis only one constant r_{11} is necessary to relate a change in index of refraction or an induced birefringence to the applied field. The new indices are given in eqn. (1):

$$n^2 = \frac{1}{O^2 \pm r_{11}E} \tag{1}$$

Here O corresponds to the inverse of the original index as can be seen by setting either $E=0$ or $r_{11}=0$. In general $r_{11}E$ is much less than O^2 and it is seen that the two new indices correspond to a slight shift, splitting symmetrically about the original index. The effect of this induced birefringence is to create a relative phase shift between two waves polarized in these two directions. Equation (2) relates δ_0 to E:

$$\delta_0 = \frac{2\pi n_0^3 r_{11} E}{\lambda} \qquad (2)$$

The effect of this phase shift on incident plane polarized light is shown in Fig. 1. The incident light is resolved along the fast and slow directions and the result of the phase-shift is to convert incident plane polarized light into generally emergent elliptical polarized light. This happens whether the birefringence is natural or induced or a combination. For light traveling down the c-axis selenium is not birefringent and any ellipticity is caused by the electric field. The reason modulation can be accomplished is simply that the amount of light which a linear analyzer transmits depends on the ellipticity of the light which can be controlled by an electric field. It is possible to achieve complete extinction by producing a relative phase shift of $180°$.

In selenium the measurement of the Pockels constant is complicated by the presence of optical activity. The effect of optical activity on plane polarized

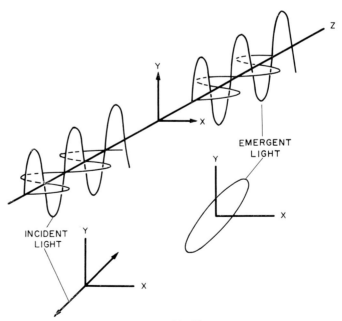

FIG. 1. Pure birefringence.

light is shown in Fig. 2. This corresponds simply to a rotation of the plane of polarization. Here there is no phase shift between components but the relative amplitudes change. We have measured the optical activity between 0.74 micron and 1.5 microns and the results are shown in Fig. 3. The large dispersion near the absorption edge is characteristic and complicates measurement in this region unless a monochromatic source is used. For the samples used the actual rotation at 1.15 microns varied from 19° to 35°. The physical significance of

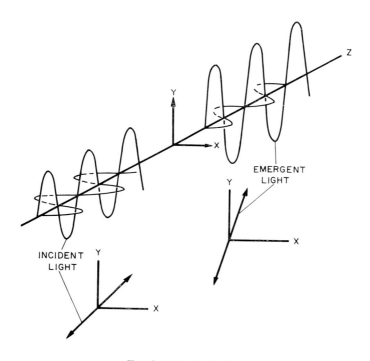

FIG. 2. Optical activity.

this rotation is that the components of the incident light in the directions of the induced birefringence are changing in amplitude continuously as the beam passes through the sample. The combined effects are shown in Fig. 4. For plane polarized incident light the emergent light is elliptically polarized and the major axis of the ellipse is rotated out of the incident plane of polarization.

The samples used were grown by R. C. Keezer of the Xerox Corporation and were aligned by X-ray techniques. Samples were prepared from slices cut from large selenium single crystal boules, with a string saw using sodium monosulfide (Na_2S) as solvent. Polishing was accomplished on a small rotating table covered with Buehler AB microcloth using first 25.0 and 5.0 micron aluminum

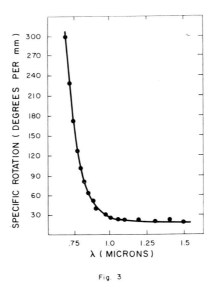

Fig. 3. Optical activity of trigonal selenium.

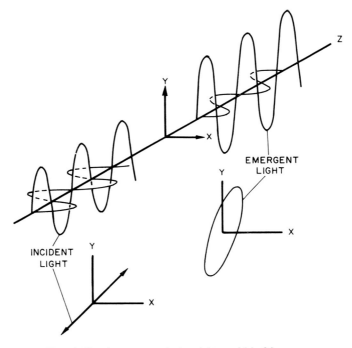

Fig. 4. Simultaneous optical activity and birefringence.

oxide. Final finish was achieved with 0.05 micron Al_2O_3. Alignment was verified by observation with the polarizing microscope of the uniaxial interference figure. For thin samples it was found that this interference figure was very convenient for alignment as a two-degree shift in sample orientation produces noticeable asymmetry in this pattern. Samples were then mounted in a goniometer on an optical bench, for fine alignment. Using a HeNe laser at 1.15 microns it was possible to achieve an extinction ratio of 60 : 1 through at 0.64 mm samples. Because of the large natural birefringence of selenium ($\Delta n = 1$), sample misalignment manifests itself in a severe deterioration of the extinction ratio and it is felt that this constitutes an excellent criterion for fine alignment. The effects of changes of a factor of 20' are easily seen. However, rough alignment from X-ray data is necessary since the extinction ratio goes through several maxima for large rotations, the one closest to the c-axis being about 3 degrees from the c-axis. The Pockels constant was determined using a crossed polarizer and analyzer configuration, compensated for optical activity with a.c. voltages at 640 cps up to 7000 volts p.p. Following a technique outlined by Sliker,[1] the electro-optic coefficient r_{11} was determined to be

$$r_{11} = 5 \times 10^{-10} \text{ cm/volt}$$

This value is higher than the measurement of Teich and Kaplan[2] at 10.6 microns.

ACKNOWLEDGMENT

The authors are indebted to Dr. James H. Becker for many helpful suggestions during the course of this work.

REFERENCES

1. SLIKER, T. R., *J. Opt. Soc. Am.* **54**, 1348 (1964).
2. TEICH, M. C. and KAPLAN, T., *IEEE J. Quantum Electronics*, October 1966.

INVESTIGATIONS ON CRYSTALLINE TELLURIUM AND SOLID AMORPHOUS AND LIQUID SELENIUM WITH INELASTIC NEUTRON SCATTERING

A. Axmann, W. Gissler and T. Springer

Institut für Festkörper- und Neutronenphysik, Kernforschungsanlage Jülich, Jülich, Germany

INTRODUCTION

So far, a number of investigations concerning the lattice dynamics of selenium and tellurium have been performed. Nevertheless, the identification of the various spectral lines and the knowledge of the vibrational modes are not yet satisfactory. Investigations with infrared spectroscopy by Caldwell and Fan[1] have shown numerous lines. These have been interpreted by theoretical calculations of Boitsov[2,3] who assumed that these lines are single transitions. Hulin[4] used a model with three force constants (bending stiffness and bonds between neighbor atoms within a spiral, and nearest neighbor forces between spirals) which have been determined by the elastic constants.[5] The results agree with the optical data partly. In this work the vibrational modes of selenium and tellurium have been determined by neutron spectroscopy. The experiments on polycrystalline and amorphous samples give the position of the maxima in the state density of the normal modes. Good agreement has been found with infrared experiments of Grosse et al.[6] and Geick.[7] A number of lines found by Caldwell and Fan[1] have been discarded. They are due to multiple processes which are easily avoided in neutron spectroscopy. Experiments on liquid selenium show that the optical modes are not essentially influenced by the melting process. Furthermore, information on the diffusive motion of the atoms can be obtained from the width of the quasi-elastic line.

GENERAL FEATURES OF THE METHOD

It seems useful to compare the essential features of neutron spectroscopy with the well-known optical methods.

(i) Because of the short wavelength of the neutrons (order of 4 Å) there is the possibility of interaction with phonons all over the Brillouin zone,

(ii) Selection rules for neutrons are different from those for optical spectroscopy; transitions forbidden in infrared spectroscopy can be observed with neutron scattering.

(iii) Multiple transitions can be avoided. The intensity of a ν-fold transition is proportional to $(Ku)^{2\nu}$ where ν is the order of multiplicity, u^2 is the average square amplitude of the scattering atom, and $\hbar K$ is the momentum transfer vector of the neutron during the scattering process

$$\mathbf{K} = \mathbf{k}_0 - \mathbf{k} \qquad (1)$$

($\hbar k_0$, $\hbar k$ = momentum of the neutron before and after the scattering process, respectively). This quantity can be chosen sufficiently small by proper experimental conditions.

For one-phonon processes, energy conservation gives

$$E_0 - E = \hbar \omega_i(\mathbf{q}) \qquad (2)$$

$\hbar \omega_i$ are the energies of the phonons with wave number q which have been excited or de-excited by the neutron ($i = 1$ to 9 for a three atomic lattice), E_0, and E are the neutron energies before and after scattering, respectively. For coherent scattering (as in the cases of tellurium, selenium, and most other elements) there holds the additional condition of momentum conservation between neutron, phonon and the lattice

$$\mathbf{k}_0 - \mathbf{k} = \mathbf{K} = 2\pi\boldsymbol{\tau} - \mathbf{q}(\omega) \qquad (3)$$

$\boldsymbol{\tau}$ is a reciprocal lattice vector. Combination of both equations allows scattering only for discrete neutron energies, E. This fact has been very broadly used for the determination of dispersion branches $\omega_i(\mathbf{q})$ in single crystals.

For a polycrystal, the situation is complicated because $\boldsymbol{\tau}$ takes all orientations in space relative to $\mathbf{K}$, and a continuous distribution rather than discrete phonon lines is obtained. Nevertheless, one finds valuable information because extrema of the functions $\omega_i(\mathbf{q})$ in reciprocal space are responsible for singularities in the density of states (critical points).[8] They can be observed as peaks in the scattered spectrum. Equation (3) introduces only a certain restriction for the scattering process. The situation is shown schematically in Fig. 1. The thin lines represent the relation between energy transfer $\hbar\omega$ and momentum transfer $\hbar K$ for a fixed scattering angle 2θ. The shaded regions are bordered by the curves

$$2\pi\tau \pm q(\omega) \qquad (4)$$

Scattering is only possible where the shaded regions are crossed by the lines $2\theta =$ const. Qualitatively, the situation is quite similar for the amorphous and liquid state if one allows for a certain width of the reciprocal lattice vectors. The behavior at very small values of ω in the liquid case will be discussed in

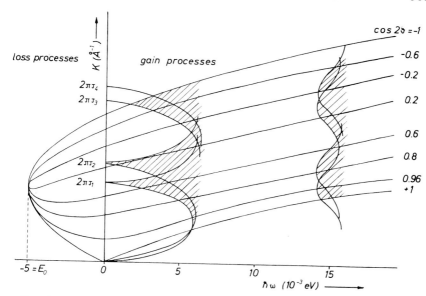

FIG. 1. (K, ω)-plane. Thin lines: relation between momentum transfer $\hbar K$, and energy transfer $\hbar\omega$ for a given scattering angle. Shaded regions are covered by equation (4) in a polycrystal (see text). τ_i = reciprocal lattice vectors. The dispersion branches are drawn only qualitatively.

Section 4. (For a detailed review on the aspects and methods of neutron spectroscopy see ref. 9; a short review has been published in ref. 10.)

EXPERIMENTAL EQUIPMENT

The energy distribution of the scattered neutrons has been measured with a rotating crystal spectrometer.[11] It is shown in Fig. 2. A subthermal neutron beam of the FRJ-2 (DIDO) reactor hits an aluminum single crystal. It reflects a monochromatic beam with an energy of 0.005 eV and width of 0.0002 eV. For the time-of-flight analysis after scattering, pulses are produced by rotating the crystal around its reflecting planes. When the beam sweeps over the collimator, a 40-microsecond burst of scattered neutrons leaves the sample. Their energy distribution is measured by time-of-flight analysis using a multichannel analyzer. The energy resolution width is 8 per cent at $E = 0.025$ eV and 6 per cent at $E = 0.005$ eV.

During the experiments the following error sources had to be investigated:
(i) Background, mainly from sample containers. It produces no structure in the spectrum and can be easily subtracted.
(ii) Frame overlap: neutrons scattered with energy loss from a certain neutron burst are detected during the following one. By changing the

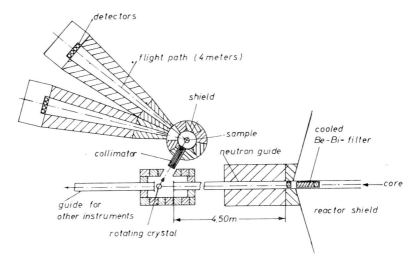

Fig. 2. Rotating crystal spectrometer for time-of-flight analysis.

rotation speed of the crystal, no change in the spectrum has been observed.
(iii) Multiple scattering of a neutron within the sample. It could create false lines (e.g. at the sum or difference of line energies). This has been suppressed by sufficiently thin samples. (Intensity ratio of double over single scattering $\sim 10^{-2}$ for a given line.)
(iv) Multi-phonon scattering which has qualitatively the same effect as (iii). Theoretical calculations using the phonon expansion[12] show that the intensity ratio of two-phonon over single-phonon lines is 4×10^{-2} or less.

RESULTS AND DISCUSSION FOR THE CRYSTALLINE PHASE OF SELENIUM AND TELLURIUM

Typical intensity distributions from selenium and tellurium at various scattering angles and sample temperatures are presented in Figs. 3 and 4, which include also the experiments on the liquid and amorphous phases. The energy transfer is $\hbar\omega = E - 0.005$ eV. The width of the elastic peak at $E_0 = 0.005$ eV is representative for the resolution of the instrument. Table 1 compares the energies of the line centers with optical measurements and calculations. *Optical modes:* The line at the highest energy (No. 5) is due to optical vibrations propagating along the c-axis of the crystal. Its position coincides actually with the average energy of the three high-energy optical branches according to Hulin's calculations[4] for tellurium (Fig. 5). The lines 3 and 4 are supposed to be due to the extremal points of the lower c-axis optical branches (see Fig. 5). If there might be a strong shift of the dispersion curves when going to other lattice directions, their extrema must be at energies lower than 0.012 eV.

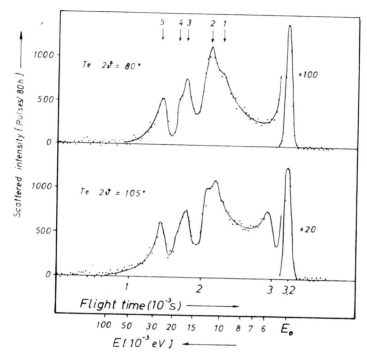

FIG. 3. Time-of-flight spectra of neutrons scattered on tellurium; abscissa: time-of-flight and energy after scattering, E. The line at $E=E_0$ corresponds to elastic scattering. $2\theta =$ scattering angle.

The levels at lower energy found with infrared spectroscopy by several authors[1, 6, 7] agree quite well with the lines 3, 4 and 5 of the present measurements (see Table 1). The higher levels given by Caldwell and Fan[1] are supposed to be due to combination lines and can be discarded. Assuming the identification of the lines as stated above there is consistency with Hulin's calculations. The great similarity between the spectra of selenium and tellurium should be especially noticed.

Acoustic spectrum. In this region, more or less sharp edges should be expected at energies where the lines $2\theta =$ const. cross the borderline of the shaded regions originating at the points $2\pi\tau_1, 2\pi\tau_2, \ldots$. This edge should shift if 2θ is changed. The low energy peak for example at 105° (Fig. 3) is probably

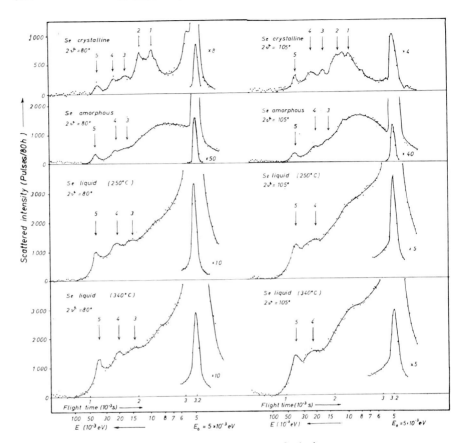

FIG. 4. Time-of-flight spectra of selenium.

due to this effect. The peaks 1 and 2 are supposed to be due to the acoustic vibrations close to the zone boundary. Because of the complexity of the situation, in this region no useful information can be obtained concerning the lattice dynamics except with single crystal studies. In the meantime such experiments have been started.

RESULTS FOR THE AMORPHOUS AND LIQUID STATE

Figure 4 shows that the optical lines 3, 4 and 5 shift only slightly when going from the crystalline to the amorphous and liquid state. This means at least that the arrangement of nearest neighbors does not change considerably and that the lowering of the binding forces is small. On the other hand, the acoustic part of the spectrum (lines 1, 2) changes completely. All structure

TABLE 1. Line Energies (in Units of 10^{-3} eV) from Neutron Spectroscopy (N), Infrared Spectroscopy (IR) and Theory (Th) (1 $cm^{-1} = 0.124 \times 10^{-3}$ eV)

Tellurium

		Line no.:	1	2	3	4	5	6	7	8	9	10	11
N		This work	4.8	6.0	10	10.5	18.5						
IR	Caldwell and Fan[1]		—	—	11	13	—	22.5	26	34	40	44	51
	Grosse et al. (6)		—	—	10.7	11.6	17.5						
	Geick[7]		—	—	—	12.5	—						
	Lucovsky et al.[16]		—	—	11.3	11.5	17.9						
Th	Hulin[4] $q=0$				9.2	15.2	20						

Selenium

			Line no.:	1	2	3	4	5	6	7	8	9
N	This work		Cryst.	4.6	7.1	11.7	16.2	31.5	—	—	—	—
			Amorph.	—	—	10.8	15.3	32.4	—	—	—	—
			Liq.	—	—	10.0	15.3	32.4	—	—	—	—
IR	Caldwell and Fan[1]		Cryst.	—	—	—	17.5	34.5	44	56	59	82
			Amorph.	—	—	—	14.5	30	45	59	75	89
	Geick and Schröder[18]		Cryst.	—	—	12.8	17.3	29.2	—	—	—	—
	Lucovsky et al.[16]		Cryst.	—	—	12.7	17.8	27.9	—	—	—	—
	Lucovsky et al.[17]		Amorph.	Many lines between 6.2 and 31.4×10^{-3} eV								

disappears. This is due to the fact that the long-range disorder disturbs the propagation of the collective acoustic modes at smaller q values.

At small energies the acoustic spectrum of the liquid merges with the elastic line which is considerably broadened. Figure 6 shows the half-width of the elastic line in selenium, after correction for the resolution width of the spectrometer, as a function of momentum transfer $\hbar K$. It can be explained by the finite time during which the oscillating atom in a liquid rests until it moves to another semi-equilibrium position. It can be shown[9, 13-15] that this width is given by

$$W = 2\hbar K^2 D \qquad (5)$$

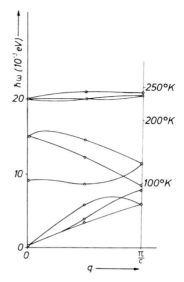

FIG. 5. Theoretical dispersion branches, $\omega(q)$, in c-direction from Hulin.[4]

where D is the self-diffusion constant of the scattering atom. In equation (5) interference effects have not been considered. They induce oscillations of W with minima at those K-values where the maxima of the diffraction pattern of the liquid are situated. Using equation (5) as an approximation for smaller K-values, one finds $D = 4.5 \times 10^{-6}$ cm²/sec at 340°C. Unfortunately there are no direct measurements of D, but it can be estimated from Eyring's formula

$$D = k_B T / 2 R \eta$$

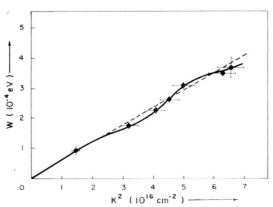

FIG. 6. Half-width of the quasi-elastic line as a function of momentum transfer $\hbar K$, on liquid selenium.

(η = viscosity). As a rough hypothesis, the effective radius R of the molecules has been assumed to be equal to the radius of the Se_6 ring-molecules (2.3 Å) which are supposed to exist above the boiling temperature. This gives $D = 1 \times 10^{-6}$ cm^2/sec which is an upper limit because R will be actually even larger because of polymerization. The reason for the discrepancy might be due to the fact that D according to equation (5) is determined by the movement of a single atom, and not of the clusters.[14, 15] Further investigations on this point might be promising.

ACKNOWLEDGMENT

The authors are grateful to Dr. Biem and Mr. Capellman for helpful discussions, and to Dr. Grosse and Dr. Geick for making available their unpublished data.

REFERENCES

1. CALDWELL, R. C. and FAN, H. F., *Phys. Rev.* **114**, 664 (1959).
2. BOITSOV, V. G., *Fiz. Tver. Tela.* **5**, 1050 (1963).
3. BOITSOV, V. G., *Fiz. Tver. Tela.* **5**, 1822 (1963).
4. HULIN, M., *Ann. Phys. (Paris)* **8**, 647 (1963).
5. BRIDGMAN, P. W., *Collected Experimental Papers*, Cambridge, Mass., Harvard University Press, 1964.
6. GROSSE, P., LUTZ, M. and RICHTER, W., *Solid State Comm.* **5**, 99 (1967).
7. GEICK, R., private communication.
8. VAN HOVE, L., *Phys. Rev.* **89**, 1189 (1953).
9. EGELSTAFF, P. A., *Thermal Neutron Scattering*, Academic Press, New York, 1965.
10. MAIER-LEIBNITZ, H. and SPRINGER, T., *Annual Review of Nuclear Science*, **16**, 207 (1966).
11. AXMANN, A. and GISSLER, W., FÜL-338NP (1965).
12. SJÖLANDER, A., *Arkiv för Fysik* **14**, 315 (1958).
13. BROCKHOUSE, B. N., *Nuova Cim. Suppl.* **9**, 45 (1958).
14. VINEYARD, G. H., *Phys. Rev.* **110**, 999 (1958).
15. LARSSON, K. E. and DAHLBORG, U., *Physica* **30**, 1561 (1964).
16. LUCOVSKY, G., KEEZER, R. C. and BURSTEIN, E., *Solid State Comm.* **5**, 439 (1967).
17. LUCOVSKY, G., MOORADIAN, A., TAYLOR, W., WRIGHT, G. B. and KEEZER, R. C., *Solid State Comm.* **5**, 113 (1967).
18. GEICK, R. and SCHRÖDER, U., *The Physics of Selenium and Tellurium*, Pergamon, New York, 1969, p. 277.

DISCUSSION

SIEMSEN: Would it be possible to use this method with a better resolution to look for the three energy bands in the area of 20 meV separately?

GISSLER: I think it is not possible for polycrystalline samples as used in our work because of anisotropy effects.

BECKER: Can you tell from your data whether or not momentum conservation is being observed for the amorphous solid selenium or liquid selenium?

GISSLER: Yes, in so far as a reciprocal lattice vector can be attributed to a liquid or amorphous substance.

THE CONTRIBUTION OF THE LATTICE VIBRATIONS TO THE OPTICAL CONSTANTS OF TELLURIUM

R. GEICK

Physikalisches Institut der Universität Freiburg, Freiburg, Germany

P. GROSSE and W. RICHTER

II. Physikalisches Institut der Universität zu Köln, Köln, Germany

INTRODUCTION

In contrast to the covalent bonded crystals silicon and germanium, the symmetry of selenium and tellurium allows infrared-active lattice vibrations, mostly due to the lack of inversion symmetry.

Microwave measurements ($\lambda = 0.8$ to 3.5 cm) by Wagner[1] have shown a remarkably higher dielectric constant than interference measurements in the near infrared ($\lambda = 4$ to 14 microns). Therefore, in tellurium strong infrared-active lattice vibrations are expected in the far infrared region with a considerable dipole moment, caused by a deformation of the valence electron shell, corresponding with an effective charge at the positions of the tellurium atoms. Because of the expected high oscillator strength, these lattice vibrations have been found as reststrahlenbands in the reflection spectrum.[2-4] Meanwhile Lucovsky *et al.*[5] have also published similar measurements. On the other hand, the early absorption measurements of Caldwell and Fan[6] yielded similar results, but did not allow for a detailed analysis.

POINT GROUP SYMMETRY AND NORMAL MODES

For infrared- and Raman-active one-phonon processes one must consider only the symmetry of the point Γ in the Brillouin zone, because the wave vector of the related phonons is $q \approx 0$. This point has the symmetry of the point group 32.

Tellurium, with three atoms per basis cell, has six optical lattice branches. Group theoretical analysis[6, 7] shows the degeneracy, the symmetry character and the related optical activity (Table 1):

(a) One single mode of symmetry A_1, Raman-active only (a "breathing" mode).

(b) Two double degenerated modes of symmetry E, Raman-active and infrared-active only for polarization $E \perp c$ (a combination of a "shearing" and a "stretching" mode), here called E' and E''.

(c) One single mode of symmetry A_2, infrared-active only for polarization $E \| c$ (a "twist" mode).

TABLE 1. *Vibrational Modes in Tellurium*

	neutron sc. meV	optical measurement			activity	
		meV	°K	cm^{-1}		
A_1					Raman	A_1
E'	18.5	17.6	204	142	Raman IR - $E \perp c$	E'
E''	10.5	11.4	133	92.0	Raman IR - $E \perp c$	E''
A_2	10.0	10.7	125	86.5	IR - $E \| c$	A_2

EXPERIMENTAL

All investigated samples were cut with a wire saw from tellurium single crystals grown by the Czochralski method (hole concentration in the impurity range *ca.* 10^{14} cm^{-3}). The surfaces were ground with corundum (grain 23 microns) and polished with diamond (grain 7 to 0.25 microns). After mechanical treatment a layer of 50 microns thickness was removed by a chemical polish (Honeywell, HCl–CrO$_3$). Reflectivity samples had a diameter of 12 to 18 mm and a thickness of 5 mm, the transmission samples a diameter of 10 to 12 mm and a thickness of 0.1 to 2 mm. There were prepared samples with [10$\bar{1}$0] cut for both polarizations and with [0001] cut for $E \perp c$ polarization only.

The optical measurements were carried out with a Perkin–Elmer spectrophotometer 301. The light was polarized by a wire grid polarizer. The samples were mounted in a glass helium cryostat with polyethylene windows.

The measurements were made in the spectral region from 20 to 200 microns and at the temperature 4°K, 90°K and 300°K.

The reflection spectra for both polarizations are shown in Fig. 1 for 300°K and in Fig. 2 for 4°K. The 90°K spectra show no additional details. In the spectra for $E \perp c$ two reststrahlenbands were found: a small one at *ca.* 70 micron wavelength and a strong one at *ca.* 100 microns; in the spectra for $E \| c$ one reststrahlenband at *ca.* 110 microns. At low temperatures all these bands become sharper.

The absorption spectra (Fig. 3) show very strong absorption peaks at the wavelengths of the reststrahlenbands. Moreover, some minor peaks are found at shorter wavelengths. These decrease weakly with decreasing temperature.

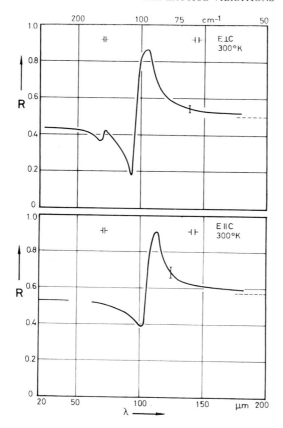

FIG. 1. Reflectivity measurements on single crystal tellurium at 300°K for polarization $E \perp c$ and $E \parallel c$. - - - - - Reflectivity calculated with the dielectric constant measured at microwave frequencies by Wagner.

ANALYSIS OF THE REFLECTION SPECTRA

The shape of the reststrahlenbands is reproduced by a classical oscillator model:

$$\epsilon = \epsilon_\infty + \frac{\Delta\epsilon \cdot \tilde{\nu}_0^2}{\tilde{\nu}_0^2 - \tilde{\nu}^2 + i\gamma\tilde{\nu}}$$

The values of the parameters are obtained by an optimum fit, namely, the value for the eigenfrequencies $\tilde{\nu}_0$, the oscillator strengths $\Delta\epsilon$ and the damping constants γ. The value for ϵ_∞ is taken from near infrared data. The results for different temperatures are listed in Table 2.

The oscillator strength $\Delta\epsilon$ is temperature independent within the limits of the experimental error, thus indicating once more that we are dealing with one-

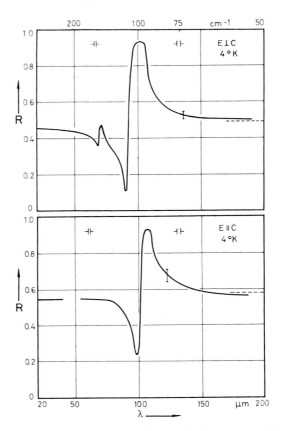

FIG. 2. Reflectivity measurements on single crystal tellurium at 4°K for polarization $E \perp c$ and $E \| c$. ----- Reflectivity calculated with the dielectric constant measured at microwave frequencies by Wagner.

phonon transitions. The contribution of the lattice vibrations to the dielectric constant, the oscillator strength $\Delta \epsilon$, explains exactly the decrease of the dielectric constant from ϵ_0 at microwave frequencies to ϵ_∞ at near infrared frequencies (Table 2). So, in pure tellurium, no other strong dispersion mechanism is expected in the infrared region.

The eigenfrequencies $\tilde{\nu}_0$ and damping constants γ show at different temperatures the usual behavior: with decreasing temperature the eigenfrequencies increase slightly and the damping constants decrease.

By comparing the experimental results with the selection rules, required by symmetry, we assign the reststrahlenband for polarization $E \| c$ to the lattice mode with symmetry A_2, and the two bands for polarization $E \perp c$ to the two E modes.

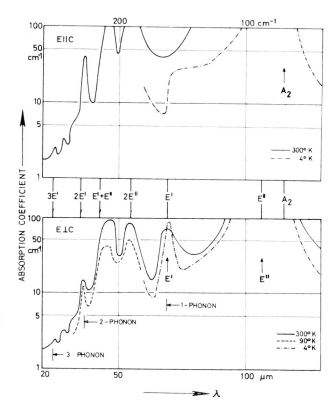

FIG. 3. Absorption spectra for polarization $E \| c$ and $E \perp c$ at different temperatures. The spectral position of combination frequencies, calculated with the eigenfrequencies at $q \approx 0$, are indicated. (Arrows: the high-energy thresholds for phonon processes of different order.)

TABLE 2. *Oscillator Parameters Deduced from Spectra*

Oscillator fit parameters: $\epsilon = \epsilon_\infty + \dfrac{\Delta\epsilon \cdot \tilde{\nu}_0^2}{\tilde{\nu}_0^2 - \tilde{\nu}^2 + i\nu\gamma}$

	$\tilde{\nu}_0/\text{cm}^{-1}$		γ/cm^{-1}		$\Delta\epsilon$	Microwave a. near IR meas. ϵ_0 : $\lambda = 0.8\text{–}3.5$ cm ϵ_∞ : $\lambda = 4\text{–}14$ μm
	300°K	4°K	300°K	4°K	300.4°K	
E'	142	143	6.5	2.3	0.73	$\Delta\epsilon = \epsilon_0 - \epsilon_\infty$
E''	92	95	4.0	1.0	11	$10 = 33 - 23$
A_2	86.5	88	4.0	2.5	18	$18 = 54 - 36$

ABSORPTION SPECTRA

In the region of the two strong reststrahlenbands (modes E'', A_2) for both polarization directions the sensitivity was not sufficient to determine the absorption coefficient quantitatively. This was possible only in the region of the mode E' ($E \perp c$) with a relatively small oscillator strength. The corresponding absorption line shows the temperature dependence expected for one-phonon processes (integrated absorption independent of temperature). At shorter wavelength all structure in the absorption spectra is due to multiphonon processes, mainly summation processes (Fig. 3).

A complete assignment of the absorption lines to distinct multiphonon processes is possible only by a detailed critical point analysis of the combined lattice vibration branches in the whole Brillouin zone. This analysis requires the knowledge of phonon dispersion curves. Such data will be available soon. Model calculations[8] and measurements of inelastic neutron scattering on single crystals[9] are in progress. The earlier model calculations of Hulin[7] were carried out only for the trigonal axis of the Brillouin zone. But the expected small dispersion of optical phonon branches, particularly of those with higher energies (E', A_1), allows an approximate localization on the wavelength scale for combinations of optical phonons by addition of the optically determined zone center frequencies.

With these assumptions the frequencies of the related phonons are known and it is possible to calculate the relative temperature dependence of the absorption[10] by means of

$$\alpha \propto 1 + \frac{1}{\exp\frac{\hbar\omega_1}{kT} - 1} + \frac{1}{\exp\frac{\hbar\omega_2}{kT} - 1}$$

Here ω_1, ω_2 are the frequencies of the phonons involved.

TABLE 3. *Relative Temperature Dependence of Absorption*

Combination	Wavelength microns	Relative temperature dependence		
		300°K	90°K	4°K
$E' + E'$	35.2	3.0	1.2	1.0
$E' + E''$	42.8	3.8	1.4	1.0
$E'' + E''$	54.4	4.6	1.6	1.0

The calculations (Table 3) yield a decrease of the absorption by a factor 2.5 to 3.0 for the combinations of the phonons E' and E'', when the temperature decreases from 300°K to 90°K. This is in good agreement with the present experimental results: the small decrease of the $E' + E'$ combination at higher energies and the greater decrease of the combinations with E'' at lower energies.

DISCUSSION

First, a comparison can be made of the present results with the data from inelastic scattering of neutrons on polycrystalline samples of tellurium. In these experiments Axmann, Gissler and Springer[9, 11] found five peaks for different energies of scattered neutrons. Three of these are associated with neutrons scattered by phonons of the optical branches. The corresponding energies are listed in Table 1. The values are in good agreement with the present results. An exact agreement is to be expected only for branches with a small dispersion, because optical experiments give the phonon energies for wave vector $q \approx 0$, while neutron scattering experiments on polycrystalline samples give particularly the phonon energies for the points of high density of states in the Brillouin zone.

In comparison to selenium[4, 5, 8, 12] the oscillator strengths of the restrahlenbands of the optically active phonons are greater in tellurium than in selenium. This means, in tellurium the valence electron shell is easier to deform in agreement with the higher polarizability of tellurium.

In tellurium the two modes E'' and A_2 have almost the same energy (at the point $q \approx 0$), while in selenium they are split widely. This agrees with the lower anisotropy of tellurium. The selenium/tellurium lattice may be considered as a cubic primitive lattice, with the [111] direction as the trigonal c-axis, deformed by a small translation of an atom in a [110] direction, so that the six nearest neighbors of the cubic lattice become the two nearest neighbors (in the same chain) and four next nearest neighbors (in adjacent chains). The related lengthening or shortening amounts in tellurium to 9 per cent and in selenium to 20 per cent of the cubic lattice constant. In the cubic lattice the energies of the A_2 mode and the lower E mode (E'') have to be degenerated. A small distortion of the cubic symmetry removes this degeneration and the energies of these modes are split. The splitting ($\Delta E/E$) amounts in tellurium to 6 per cent and in selenium 30 per cent.

In comparison to other elemental semiconductors, tellurium has an unusually high polarizability, relatively low energies of the optical phonons, some of which are connected with a dipole moment. This gives new aspects for the discussion of the semiconductor properties of tellurium.

For all transport properties one has to expect the influence of the polar optical phonons on the carrier mobility. The drifting electrons or holes polarize and deform the surrounding lattice, i.e. they move as polarons. In tellurium with a relatively low polaron coupling constant $\alpha \approx 0.3$† we have not to expect a "polaron mass" markedly higher than the normal effective carrier mass ($m_{polaron} = (1 + \alpha/6)m_{eff} \approx 1.05 m_{eff}$. But the scattering of carriers by polar optical phonons is the most important contribution to the mobility.[5]

---

† This means the lattice deformation energy of the drifting carrier measured in units of the energy $k\theta$ of the longitudinal optical phonon.[13]

To estimate the amount of optical phonon scattering, the numerical values were calculated according to the theory of Howarth and Sondheimer[13] and compared (Fig. 4) with the experimental total scattering of holes (reciprocal mobility) derived from d.c. measurements. The theory gives remarkably higher values for scattering than the experiments.

Another problem is the strong thermal excitation of lattice vibrations because of the low Debye temperature. For all properties caused by the coupling of the electron eigenfunctions of the lattice atoms, the distance of the

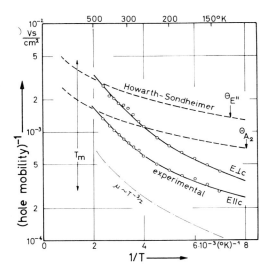

FIG. 4. Hole mobility and polar optical phonons in tellurium. - - - - Hole scattering calculated according to the theory of Howarth–Sondheimer ($m_{hole} = 0.4\ m_0$). —o—o— Experimental—d.c. measurements. —·—·— Proportional to acoustical phonon scattering. T_m melting temperature.

atoms is very important. The equilibrium positions of the atoms changes with temperature by thermal expansion because of the anharmonicity of the lattice vibrations. This expansion is in tellurium very anisotropic. But in concurrence with the thermal expansion is the thermal excitation of harmonic vibrations of the atoms around their equilibrium position, which will be anisotropic too.

These vibrations change the anisotropy of the coupling, because a short time overlap of the electron orbits of atoms, oscillating with antiphase, contributes to the coupling integral with a higher weight in the nearest position than in the widest one.

To compare the temperature dependence of thermal expansion and excitation, the relative change of the two lattice constants was estimated (Fig. 5) by calculating the expansion by the linear expansion coefficient and the excitation

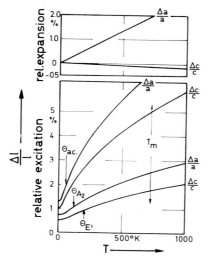

FIG. 5. Temperature dependence of lattice deformation. Relative change of the lattice constants a and c—expansion calculated by linear expansion coefficient; excitation calculated with a harmonic oscillator of different Debye temperatures.

by simple harmonic oscillators of different Debye temperatures excited according to a Bose–Einstein distribution.

Such an "effective" temperature dependent distance of the atoms should enable us to explain the strong temperature dependence of the anisotropy of the effective hole mass in tellurium or the sign inversion of the temperature coefficient of the gap energy.[14–16]

ACKNOWLEDGMENT

The authors gratefully acknowledge Prof. Dr. J. Jaumann (Köln), for his encouragement and Dr. Axmann and Dr. Gissler (Jülich) for their discussions. Thanks are also expressed to the Deutsche Forschungsgemeinschaft for assistance with equipment.

REFERENCES

1. WAGNER, H., *Z. Physik* **193**, 218 (1966).
2. GROSSE, P., LUTZ, M. and RICHTER, W., *Solid State Comm.* **5**, 99 (1967).
3. RICHTER, W. and GROSSE, P., *Verhandl. DPG* (VI) **2**, 29 (1967).
4. GEICK, R., *Verhandl. DPG* (VI) **2**, 29 (1967).
5. LUCOVSKY, G., KEEZER, R. C. and BURSTEIN, E., *Solid State Comm.* **5**, 439 (1967).
6. CALDWELL, R. S. and FAN, H. Y., *Phys. Rev.* **114**, 664 (1959).
7. HULIN, M., *Ann. Phys.* (*Paris*) **8**, 647 (1963).

8. GEICK, R. and SCHRÖDER, U., *The Physics of Selenium and Tellurium*, Pergamon, New York, 1969, p. 277.
9. AXMANN, A., GISSLER, W. and SPRINGER, T., *The Physics of Selenium and Tellurium*, Pergamon, New York, 1969, p. 299.
10. JOHNSON, F. A., *Proc. Int. School of Physics "Enrico Fermi", Course XXII, Semiconductors*, Academic Press, New York, 1963, p. 504.
11. AXMANN, A. and GISSLER, W., *Phys. Status Solidi* **19**, 721 (1967).
12. GEICK, R., SCHRÖDER, U. and STUKE, J., *Phys. Status Solidi* **24**, 99 (1967).
13. EGGERT, H., *Festkörperprobleme*-Band I, Editor F. SAUTER, Braunschweig, Friedr. Vieweg & Sohn, 1962, p. 295.
14. GROSSE, P. and LUTZ, M., *Verhandl. DPG* (6) **2**, 30 (1967).
15. GROSSE, P. and SELDERS, M., *Phys. Status Solidi* **24**, K33 (1967).
16. GROSSE, P., *Die Festkörpereigenschaften von Tellur;* Springer Tracts in Modern Physics, Vol. 48, Heidelberg/New York, Springer, 1969.

ELECTRICAL PROPERTIES

INVESTIGATION OF THE CHARGE-TRANSFER STATE IN SELENIUM BY THE ELECTRONIC PARAMAGNETIC RESONANCE METHOD

G. B. ABDULLAYEV, N. I. IBRAGIMOV and SH. V. MAMEDOV

Institute of Physics, Academy of Sciences of the Azerbaidzhan, Baku, U.S.S.R.

INTRODUCTION

Among semiconductors selenium finds the widest applications and in some of these it is irreplaceable. In almost all of the properties of selenium, however, anomalies are observed. These anomalies are probably connected with the donor–acceptor interactions with the participation of such strong acceptors as oxygen, iodine, etc. The electronic paramagnetic resonance investigations of the interactions of such give significant information about these anomalies allowing us to check the properties of selenium.

A widely held opinion exists that the impurities in selenium are found in the form of a compound. Experiments on the investigation of the state of impurities by the electronic paramagnetic resonance method are, however, convincing us that the predominant role is played not by such compounds but by the donor–acceptor interactions with charge-transfer. The halogens hold a special place as dopants and adequate consideration has not been given so far to the active state of oxygen in selenium. Below are given the results of an investigation of the interaction of selenium with the acceptor impurities, oxygen and iodine.

EXPERIMENTAL

Measurements were carried out on a radiospectrometer with double modulation of the magnetic field. The operating frequency was equal to 9304 Mc/s. The standards used were DPPH and Mn^{2+} in ZnS (natural sphalerite).

As a basic parent material amorphous powdered selenium of 99.9999 per cent purity (B-5) was used, but the investigations were also extended to selenium of lesser purity. From these materials samples with different mole content of oxygen and iodine (0.005, 0.01, 0.03, 0.05, 0.1, 0.3, 0.6 and 1 per cent of I_2) were prepared. Samples of two series were prepared: (a) directly from the amorphous powder and (b) from selenium first fused, crystallized and ground in a mortar. The size of grains was equal to ~ 50 microns. The iodine impurities were introduced in a definite mole proportion into the powder, and then the ampoules were evacuated to 10^{-5} mm of Hg. In so doing, a part of the iodine introduced into the selenium was also evacuated. Taking this into consideration, the suspension of iodine was increased by 20 per cent. The oxygen impurities were also introduced by means of a partial cock into powdered selenium. Before the introduction of oxygen the ampoules were evacuated up to 10^{-5} mm of Hg and then filled with oxygen in the range of 10^{-1}–10^{-5} mm of Hg.

The electronic paramagnetic resonance absorption was investigated in all the samples without any heat treatment. A weak signal was observed only in those samples with 0.6 and 1 mol. per cent of I_2 prepared from the sieved powder. Later on each sample was exposed to a successive heat treatment for 2 hr in the interval of 180° to 700°C. After each stage of heat treatment the electronic paramagnetic resonance absorption was investigated.

RESULTS

In the samples containing oxygen as well as iodine an almost symmetrical singlet without any hyperfine structure was observed, with $g = 2.0035 \pm 0.0005$ and with the width between the points of maximal slope (peak-peak width) $\Delta H_m = 5.5 \pm 0.5$ Oe. A typical curve of absorption is shown in Fig. 1. In the interval $-196°$ to 17°C the ΔH_m and g values of the resonance line did not change, and the increase of intensity of the signal was observed in accordance with Curie's law. The intensity of the signal depends on the concentration of the impurity as well as on the regime of heat treatment of the sample. The diagrams of the intensity change with the change of T_{treat} are shown in Figs. 2 and 3, the intensity being expressed in relative units (I/I_{st}), where I_{st} is the intensity of the standard signal. As is seen from the diagrams, the intensity of the electronic paramagnetic resonance absorption first increases with the increase of T_{treat}, reaches a maximum, and then sharply decreases until the disappearance of the signal at a definite temperature. Depending on the concentration of impurity, the formation, the passing through maximum, and the disappearance of the signal, take place at different values of T_{treat}. As a rule, at large concentrations the signal originates at small T_{treat}, and at the concentrations > 0.6 mol. per cent of I_2 in the sieved samples a weak signal originates even without heat treatment. It is necessary to mention that the time of heat treatment was always equal to 2 hr, and the values of T_{treat} were held constant to $\pm 3°$ by means of the thermoregulator.

Figure 2 shows that at the greater iodine content the general shape of the temperature dependence of the intensity has considerable anomalies, although

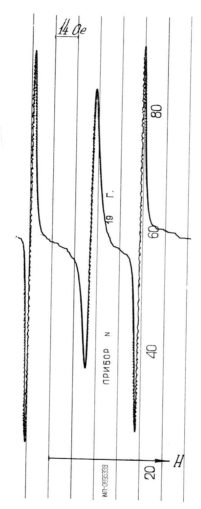

Fig. 1. Spectrum of the electronic paramagnetic resonance absorption in selenium (central line). Lines from the right and left sides relate to Mn^{2+} in ZnS (natural sphalerite).

some correlation is retained. This evidently is due to the fact that at the large percentage contents of impurities (0.30, 0.60 and 1 mol per cent of I_2) the iodine does not dissolve in the selenium and there occurs a separation of a second phase.

For this reason analysis of the results of the experiments was carried out without taking into consideration data of samples with iodine content over 0.3 mol per cent of I_2. In selenium with impurities of oxygen and iodine in the

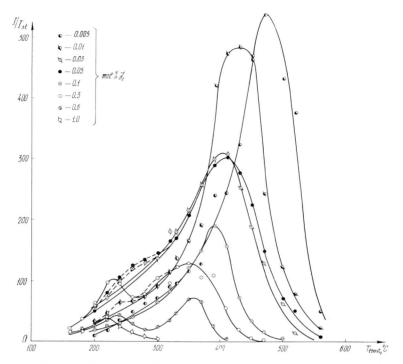

FIG. 2. Intensity dependence of the electronic paramagnetic resonance signal in selenium with the impurities of I_2 on the temperature of heat treatment (T_{treat}).

range of $T_{\text{treat}} \sim 300\,°\text{C}$ in the diagram of temperature dependence of the intensity, a characteristic convexity is revealed (dashed line in Figs. 2 and 3). This is an indication of the process being associated with large concentrations of the acceptor.

The temperature of the intensity maximum does not depend on the modification of the parent selenium, since in both groups of samples a similar behavior is observed.

In the case of oxygen admixture the temperature dependence of the intensity is of the same character (Fig. 3).

It was reported previously[1] that in selenium of < 99.999 per cent purity a signal was not discovered. Later investigations showed that by carrying out the appropriate procedure of heat treatment, the electronic paramagnetic resonance absorption was observed also in selenium of lesser purity. Figure 3 shows the dependence of the intensity on T_{treat} for selenium of 99.996 per cent purity, in which neither the nature of the impurities nor the pre-history is known. In these samples even without pulverizing (i.e. in the bulk samples), a noticeable electronic paramagnetic resonance absorption was observed.

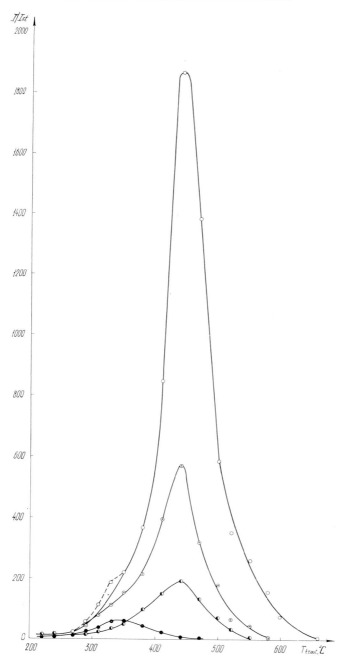

FIG. 3. Intensity dependence of the electronic paramagnetic resonance in selenium with the impurities of O_2 on the temperature of heat treatment (T_{treat}):

○   selenium B-2, crushed, $d \sim 50\mu$, $p = 10^{-4}$ mm Hg;

◐   selenium B-5, $p = 10^{-3}$ mm Hg;

◌   selenium B-2, bulk sample, $p = 10^{-4}$ mm Hg;

●   selenium B-2, $p = 760$ mm Hg.

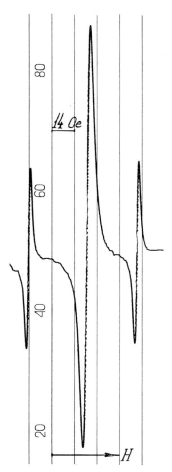

FIG. 4. Spectrum of the electronic paramagnetic resonance absorption in the crushed selenium B-2, $T_{\text{treat}} = 400°C$. $t = 2$ hr, $d \sim 50$ μ, $p = 10^{-4}$ mm Hg.

Pulverizing leads to a considerable increase of the intensity (Fig. 4). The maximum concentration deduced was $\sim 10^{17}$ cm^{-3}.

On the basis of the data of Figs. 2 and 3 the diagrams of dependence

$$\log (I/I_{\text{st}}) = f\left(\frac{1}{T_{\text{treat}}}\right)$$

have been plotted. This dependence is well demonstrated in Figs. 5 and 6 by the straight lines, the inclination of which characterizes the energy necessary for the formation of the paramagnetic center and depends on the concentration and the nature of the impurity introduced.

THE CHARGE-TRANSFER STATE IN SELENIUM 327

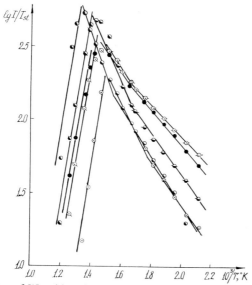

FIG. 5. Variation of I/I_{st} with reciprocal temperature for the samples given in Fig. 2.

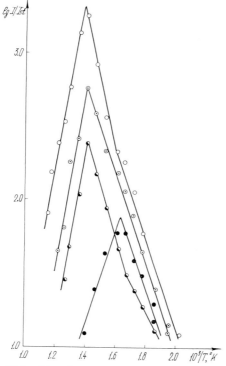

FIG. 6. Variation of I/I_{st} with reciprocal temperature for the samples given in Fig. 3.

In the previous work[2] it was noticed that the most probable source of signal of the electronic paramagnetic resonance in selenium was the charge-transfer complex. This supposition has found its confirmation in the case of iodine impurity as well.

DISCUSSION

Let us consider the question of the process leading to the formation of the charge-transfer complex.

The presence of the two molecules of the excited states of the system connected with the electron transfer from one component to another was for the first time predicted by Mulliken[3] and developed by Terenin,[4] Blumenfeld and others.[5-8]

The following mechanism of formation of the charge-transfer complex in selenium with the participation of the acceptor molecules of O_2 and I_2 is now presented. As is known, atoms in the selenium chain have a covalent bond with the interchain bond being of the Van der Waal's type. At the ends of the chains of selenium there exist non-saturated bonds characterized by the fact that the two p-electrons of the end atoms of the chain do not take part in the linkage.

This situation also applies for amorphous selenium, as the mechanism of formation of the charge-transfer complex for the amorphous as well as for the crystalline selenium is similar. An attempt to reveal any change in the parameters of the signal by transformation of selenium from the amorphous into the crystalline modification did not lead to any positive result, probably the localized structure giving rise to the signal in the amorphous sample is completely preserved at crystallization. The ends of the selenium chains are the centers of absorption, which is why the impurity molecules initially are physically bound to the ends of the chains. During the subsequent heat treatment conditions are created which drag the electron out of the chain-end into the molecular orbit of the acceptor. Thus the complex with the charge-transfer is formed. This is stimulated by the large affinity for the electron of O_2 (0.89)[9] and of I_2 (1.8 eV)[10] molecules, and also by high polarization, and probably, by the comparatively low ionization potential of the selenium chain. Unfortunately there is no information concerning the ionization potential of long chains of selenium in the literature. The available data on the ionization potential of separate atoms cannot be taken into consideration as this is considerably different from the ionization of a long chain. However, it may be affirmed that the ionization energy of the chain must be essentially less than that of the atom. Thus, the acceptor molecule in the process of thermoactivation transfers from the physically absorbed state into the charge-transfer state.

The dependence of intensity on the values of T_{treat} indicates the presence of the activation process in the formation of the charge-transfer complex. From the diagrams of Figs. 5 and 6 the activation energies of formation of the charge-transfer complex at different acceptor concentrations have been determined. It turned out on the whole that at all the concentrations of iodine, a charge-transfer complex of the two types is formed: one with lesser energy of formation (type I) and one with greater energy of formation (type II). The formation energy of these two types of charge-transfer complexes depends on the concentration of acceptor. This dependence is shown in Fig. 7. With the concentration of 0.03 mol per cent of I_2 the activation energies of the charge-transfer

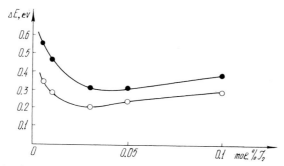

FIG. 7. Activation energy dependence of the charge transfer complex on the concentration of iodine:
○   complex with the charge-transfer of the first type;
●   complex with the charge-transfer of the second type.

complexes of the two types pass through a minimum equal to 0.21 and 0.31 eV, respectively. With further increase of concentration the activation energies increase slightly. The first type of charge-transfer complex is formed at a considerably lower value of T_{treat} and is the result of interactions of the acceptor molecules with long chains of selenium. Actually, with the increase of the length of the chain the ionization energy will decrease and the process of formation of the charge-transfer complex must proceed comparatively easily. In Fig. 5 it is seen that for all the lines the break is observed at values of T_{treat} from 330° to 350°C. This point characterizes the beginning of formation of the charge-transfer complexes of the second type with participation of more short chains of selenium.

It is known[11] that in the process of the fusing of selenium the weak interchain bonds first break down but the chain structure itself is, to a great extent, preserved. Therefore it is natural to suppose that at temperatures of 300–350°C long chains decompose into shorter ones leading to an increase of the proportion of short chains. Consequently, this leads to an increase in the probability of formation of charge-transfer complexes of the second type.

Division of the charge-transfer complexes into the two types is presented somewhat tentatively. Because of the large variety of chains of different lengths, it is difficult to consider large numbers of possible interactions of them with the acceptor molecules.

The further increase of values of T_{treat} above 400°C leads to the beginning of an irreversible decay of all the charge-transfer complexes. Beyond the dependence on acceptor concentration (excluding large concentrations), the energy of disappearance of the charge-transfer complexes is constant and equal to approximately 1.2 to 1.3 eV. At temperatures above 500°C the chain structure of selenium breaks down completely, a decay of the charge-transfer complexes and, apparently, chemical bonding of the acceptor with selenium takes place. As a result of this, the activity of the acceptor impurities falls off.

The nature of formation of the paramagnetic center in the case of the oxygen impurities is similar. For the B-5 selenium, charge-transfer complexes of the two types are again formed with activation energies $\Delta E_I = 0.4$ eV and $\Delta E_{II} = 0.7$ eV. However, in this selenium an interesting phenomenon is observed. If the sample was not subjected to pulverizing (i.e. a solid amorphous piece) then charge-transfer complexes of only the one type with $\Delta E = 0.6$ eV are formed. On the other hand, in the case of the pulverized sample, charge-transfer complexes of the two types with $\Delta E_I = 0.62$ eV and $\Delta E_{II} = 0.96$ eV are formed. The decay energy of both types of charge-transfer complexes with oxygen in selenium of different purity also appeared to be equal to 1.25 eV. As in the case of the impurities of I_2, at large concentrations of O_2 the signal appears and disappears at comparatively low values of T_{treat} and has a small intensity in the maximum. The characteristic convexity at a T_{treat} value of 300°C is also observed in this case. All this is well illustrated by Fig. 3. It is to be noted that in a similar way to I_2, large concentrations of O_2 depress the signal. As found previously,[2] at very small concentrations a signal is unobservable. Apparently the optimum pressure of O_2 is 10^{-3} to 10^{-4} mm of Hg for selenium samples B-2 and B-3. Concerning selenium B-4 and B-5, with the increase of the degree of pumping out (beginning with 10^{-1} mm of Hg), the intensity of signal is decreased (other conditions being equal). At a given stage it is difficult to take into account all the factors affecting the process of formation of the paramagnetic center in selenium. This is indicated by the complexity of the parent material and by the quantity of impurities present in it.

Recently an article was published[12] where the electronic paramagnetic resonance absorption in selenium was also observed. Along with the narrow line width with parameters coinciding with ours, a broad line was reported with the parameters $g = 2.3$ to 2.8, and $\Delta H_m \simeq 3$ kOe. This signal is attributed to selenium by the author, who explained such a large width by homogeneous and non-homogeneous broadening at the expense of the interaction of the chain-ends of selenium with one another. This supposition is partly confirmed by theoretical estimates[13] carried out by the molecular orbital method on the basis of the model of interaction of neighboring chain-ends.

In spite of a thorough search for such a wide line spectrum, we have not succeeded in finding it.

It was often observed as a wide asymmetrical line with $g \simeq 2.01$ and $\Delta H_m \simeq 200$ Oe. Control experiments showed that this line is associated with the material of the ampoule.

In studies[14-16] of measurements of static susceptibility of selenium, it was concluded that the number of chains increased with increase of temperature, i.e. the paramagnetic contribution to the susceptibility increased with temperature. We tried to observe this increase of paramagnetism by means of investigation of electronic paramagnetic resonance absorption in selenium in the temperature interval up to 800°C. A furnace of special construction was used which was installed into the resonator,[17] leaving the high quality of the resonator almost unchanged.

The high-temperature experiments have not proved the supposed increase of the paramagnetism. Only a decrease of intensity of the signal to its eventual disappearance was observed.

From this fact it was concluded that, firstly, the increase of the paramagnetic contribution reported[14-16] is apparently not connected with selenium but has another source. Secondly, the electronic paramagnetic resonance signal is not produced directly by the ends of chains of selenium, but it originates from the donor–acceptor interaction. It appears that the thermodynamical theory[18] cannot be extrapolated into the region of high temperature, although this has been done by some authors.

The explanation of the formation of electronic paramagnetic resonance absorption in selenium by the existence of unsaturated links of the ends of chains, recently proposed by us, seemed to be satisfactory. However, contradictions arose of the following type. In the parent pure material as a rule a signal is not observed whether selenium is in the form of a powder or a solid piece, although the presence of chains cannot be questioned. In all the samples, however, signals originated only in the presence of an acceptor impurity, a larger surface to volume ratio and also after heat treatment. That is why the charge-transfer complex model now appears to be the most convincing one.

A schematic model for the chemisorbed molecule of oxygen on the surface of hexagonal selenium given in the work of Sampath[12] is possible but scarcely probable because it is difficult to conceive such a geometry for the surface of selenium single crystal when all the neighboring chains terminate at least in pairs on the surface plane. Such a model to a certain extent contradicts the increase of intensity of signal with increase of the values of T_{treat}.

In fact, with the increase of T_{treat} the process of chemisorption is intensified and according to this model must lead to a decrease of intensity to account for the bonding of the ends.

In this respect a simpler and more suitable model is as follows in which $\delta \simeq 1$:

$$^{-\delta}O_2^{+\delta}.Se=Se=Se\ldots=Se=Se^{+\delta}.O_2^{-\delta}.$$

The probability of the idea of formation of the charge transfer complex is first corroborated by the numerous investigations on highly molecular compounds and also by the fact that the substitution of the acceptor molecule of oxygen by the acceptor molecule of iodine leads to a similar physical picture. Support for the model chosen of a charge transfer complex is the theoretical computation of the g-displacement carried out in reference (13) for the case of an oxygen atom at the end of the selenium chain. Not without interest are computations of the g-displacement by the model with the oxygen molecule at the ends of the chain. On this basis there is no ground not to expect the best agreement between the observed and computed values.

It is necessary to discuss in some detail the effect of using crushed material. As was mentioned above,[2] pulverizing in the presence of an acceptor in the selenium has an essential influence on the intensity of the electronic paramagnetic resonance signal because of the increase of the ratio of surface to volume. Along with this, by the process of pulverizing, local warming up of the material takes place because of the friction. As a result a transformation of the selenium may take place from one modification into another.

From another point of view, local warming up produces partial activation. As confirmation of this, there is the appearance of a weak signal after pulverizing the selenium having a larger content of iodine (~ 1 per cent) even without proper heat treatment. Thus, crushing to smaller and smaller size of particles increases the active surface of the material, and from the other aspect it intensifies the process of partial activation.

It might be supposed that the signal originates from some impurities introduced during the process of pulverizing. If this was so, then we must observe a signal in all cases independent of the oxygen or iodine content, with a temperature dependence of intensity also independent of their concentrations. The influence of foreign impurities on the intensity, however, is certainly impossible to deny.

As was mentioned, extremely large concentrations of acceptors depress the signal. This could be explained by the "oxygen" effect but the width and the g of the signal are not changed with the concentration. Only the intensity is changed. Therefore, the nature of this phenomenon is different and not yet understood.

Finally, the signal restoration is discussed. After the disappearance of the signal during the heat treatment ($T_{\text{treat}} \simeq 600\,°C$, 2 hr) partial restoration is sometimes observed by repeated treatment at $T_{\text{treat}} \simeq 400\,°C$. Control experiments in special furnaces showed that the partial restoration of the signal is connected with the presence of the temperature gradient. Probably charge-transfer complexes are created by molecular oxygen remaining unreacted in the ampoule and released in the process of the decay of the charge-transfer complex.

As a rule, the width of signal is slightly increased (10–15 Oe) at the beginning of the process of decrease of intensity (Fig. 8). ΔH_m of the partially restored signal is also about 15 Oe. The process of partial restoration also depends on the concentration of the acceptors.

The impossibility of partial restoration of the signal in the samples continuously treated at $T_{treat} \geqslant 600\,°C$ indicates the fact that, seemingly, an irreversible chemical linkage of the acceptor with selenium takes place.

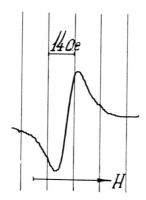

FIG. 8. Spectrum of the electronic paramagnetic resonance of selenium B-2, crushed. $d \sim 50\,\mu$, $p = 10^{-4}$ mm Hg, $T_{treat} = 580\,°C$, $t = 2$ hr.

The local paramagnetic centers due to oxygen and iodine are of extrinsic origin in regard to the basic structure of selenium. From this it is possible to state that the conductivity of selenium with a definite impurity content treated at $T_{treat} < T_m$ must be essentially larger than that of selenium treated at $T_{treat} > T_m$, where T_m is the temperature of heat treatment at which the intensity reaches its maximum value. Measurement of the electrical conductivity of the crystallized samples confirms this supposition with σ changing from 10^{-4} to 10^{-7} ohm^{-1} cm^{-1}. The electrical conductance of selenium increases with the increase of concentration of paramagnetic centers and it is possible to speak about the existence of a correlation between the conductance and paramagnetism. The measurements of the thermoelectric power showed that for all the samples it has a positive sign and is $\sim 1000\,\mu V/°C$. As it is known, the thermoelectric power characterizes the property of the grains themselves. However, the origin of the paramagnetic centers and that of the current carriers takes place at the ends of selenium chains, i.e. at the surface of the grains. The weak dependence of the thermoelectric power on the oxygen and iodine content suggests the supposition about the decisive role of chain ends in the process of paramagnetism and conductivity. The charge-transfer complexes act as if "bridges" exist, making easier the transition of the carrier from one chain into another.

Experimental data have made it possible to make one more conclusion. The process of deoxidation of selenium probably consists of the conversion of physically bound oxygen into the chemically active form with the subsequent evaporation of the oxides from the great bulk of the material.

REFERENCES

1. ABDULLAYEV, G. B., IBRAGIMOV, N. I., MAMEDOV, SH. V., DZHUVARLY, T. CH. and ALIYEV, G. M., *Doklady Akad. Nauk Azerb. SSR* **20**, 13 (1964).
2. ABDULLAYEV, G. B., IBRAGIMOV, N. I., MAMEDOV, SH. V. and DZHUVARLY, T. CH., *Phys. Status Solidi* **16**, K113 (1966).
3. MULLIKEN, R. S., *J. Am. Chem. Soc.* **74**, 811 (1952).
4. TERENIN, A. N., *Uspekhi Khim.* **24**, 121 (1955).
5. BLUMENFELD, L. A., VOYEVODSKY, V. V. and SEMENOV, A. G., *Application of the Electronic Paramagnetic Resonance in Chemistry*, SO Acad. Sci. of the USSR, N (1962).
6. BLUMENFELD, L. A. and BENDERSKY, V. A., *J. Struct. Chem. (USSR)* **4**, 405 (1963).
7. BENDERSKY, V. A. and BLUMENFELD, L. A., *Doklady Akad. Nauk SSSR* **144**, 813 (1962).
8. BLUMENFELD, L. A., BENDERSKY, V. A. and STUNZHOS, P. A., *J. Struct. Chem. (USSR)* **7**, 686 (1966).
9. MASHCHENKO, A. I., SHARAPOV, V. M., KAZANSKY, V. B. and KISELEV, V. G., *Zh. Teor. i Eksperim. Khim. Akad. Nauk Ukr. SSR* **1**, 381 (1965).
10. BRIEGLEB, G., *Angew. Chem.* **76**, 326 (1964).
11. REGEL, A. R., *Structure and Physical Properties of Matter in the Liquid State*, Edit. of Kiev State Univers., 111 (1954).
12. SAMPATH, P. I., *J. Chem. Phys.* **45**, 3519 (1966).
13. CHEN, I., *J. Chem. Phys.* **45**, 3536 (1966).
14. MASSEN, C. H., WEIJTS, L. M. and POULIS, J. A., *Trans. Far. Soc.* **60**, 317 (1964).
15. TOBISAWA, SHOTARO, *Bull. Chem. Soc. Japan* **33**, 889 (1960).
16. BUSCH, G. and VOGT, O., *Helv. Phys. Acta* **30**, 224 (1957).
17. CHAIKIN, A. M., *Pribory i Tekhnika Eksperimenta* **6**, 178 (1963).
18. EISENBERG, A. and TOBOLSKY, A. V., *J. Polymer Sci.* **46**, 19 (1960).

THE CONDUCTIVITY MECHANISM IN MONOCRYSTALLINE SELENIUM

T. Salo, T. Stubb and E. Suosara

Electronics Laboratory, Technical University of Helsinki, Helsinki, Finland

The electrical conductivity mechanism in monocrystalline selenium is not known. From the literature[1,2] one can establish that the conductivity is 10^{-5} to 10^{-6} $(\Omega\text{cm})^{-1}$ and its temperature dependence can be written

$$\sigma = \sigma_0 \, e^{-\Delta E/kT},$$

where ΔE is the activation energy. Through thermopower measurements monocrystalline selenium is found to be *p*-type with a hole concentration 10^{14} to 10^{15} cm^{-3}, independent of the temperature.[1,3] The hole mobility at room temperature has been determined to be 20 cm^2 V^{-1} sec^{-1}.[4]

The electrical conductivity σ depends on the crystal direction. Parallel to the *c*-axis the conductivity is greater than perpendicular to it.[2,3] The ratio $\sigma_\parallel / \sigma_\perp$ is found to be between the values 3 and 10. On the other hand, the activation energy is considered to be equal in both directions. All this information should be accepted with reservation. The conductivity in monocrystalline selenium is also assumed to depend on the frequency.[5]

In order to elucidate the conductivity mechanism in selenium the following measurements were made:

1. Measurements of space charge limited currents in monocrystalline selenium.
2. Measurements of the frequency and temperature dependence of the conductivity and the dielectric constant.
3. Investigation of dislocations. (These measurements are in progress.)
4. Investigation of the Hall effect with d.c. and a.c. voltage. (These measurements are in progress.)
5. Measurements of the temperature dependence of the conductivity and the dielectric constant at 24 GHz.

All measurements were carried out on samples taken from the same crystal. The impurity content of the samples was approximately 30 ppm. By means of X-ray investigations a mosaic structure in parts of the crystals was found. The dislocation studies have shown traces which can be considered to be low angle boundaries. It is evident that the conductivity mechanism should be explained with these imperfections taken into consideration.

Because the crystal contacts have a great influence on the measurements and as the present method differs to some extent from the ordinary, it will be described as follows:

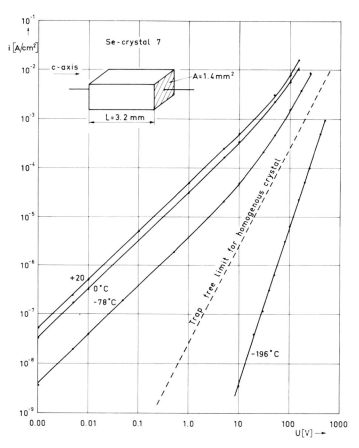

FIG. 1. Current density plotted against the voltage parallel to the c-axis at different temperatures. $L = 3.2$ mm.

After polishing with diamond paste (0.3 μ) the crystal is etched with a solution of 5 ml H_2SO_4 plus some drops of HNO_3 until the deformed layer resulting from the polishing has vanished. Then the crystal is purified by means of ion bombardment in a nitrogen atmosphere of 10–20 μm Hg after which the pressure in the vacuum-chamber is pumped down to less than 10^{-6} torr and gold contacts are evaporated onto the crystals. In this way ohmic contacts have been produced which are of great importance when measuring space charge limited currents.

By measuring space charge limited currents in monocrystalline selenium, it can be determined whether the defect concentration is homogeneously distributed.

In these crystals the current density (i) has been determined as a function of the voltage (V) at different temperatures in the direction of the c-axis. As shown in Fig. 1 the curves bend upwards at higher voltages, due to the space charge limited currents. The expression for such currents in a homogeneous crystal is

$$i = \frac{9}{8} \epsilon_r \epsilon_0 \mu \frac{V^2}{L^3}$$

where

i = the current density,
ϵ_r = the relative dielectric constant,
ϵ_0 = the dielectric constant in vacuum,
μ = the mobility of the charge carriers,
V = the voltage over the crystal,
L = the length of the crystal in the current direction.

If the greatest values existing in the literature are inserted for ϵ_r and μ, i.e. $\epsilon_r = 20$ and $\mu = 400$ cm^2 V^{-1} sec^{-1}, one obtains the curve for the trap-free limit for homogeneous crystals. With lower values for ϵ_r and μ and considering the traps in the crystal, one gets a lower curve. The measured curves are as a rule found to lie above the stated limit.

Figure 2 shows the measured potential distribution in one of the studies

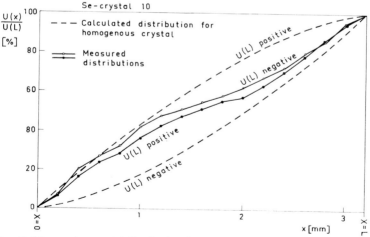

FIG. 2. Measured potential distribution in a selenium crystal compared with the calculated potential distribution for a homogeneous p-type crystal.

compared with the calculated potential distribution for a homogeneous p-type crystal. It can be concluded from these measurements that the curves do not correspond with space charge limited currents in homogeneous crystals. Consequently, the crystal embodies imperfection regions and the conductivity varies in different parts of the crystal.

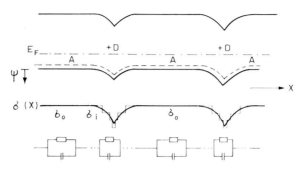

FIG. 3. The association of potential barriers with layers having different conductivities. A is ionized acceptor state, D is ionized donor state, and conductivity follows eqn. (1).

The measurements which have been carried out can be explained with a layer model,[1, 6] which is founded upon potential barriers. As much as $2°$ difference has been observed in the direction of the mosaic blocks in the crystals studied. Small angle boundaries like this cause acceptor levels in germanium and donor levels in silicon. As the forbidden band gap in selenium is greater than that in silicon, donor levels may appear in selenium. These form potential barriers in conjunction with impurities that nucleate in the small angle boundaries.

For the potential barriers in Fig. 3 the conductivity is

$$\sigma(x) = q\mu p_0 \exp\left(-\frac{q\psi}{kT}\right) \quad (1)$$

if the free path of holes is much smaller than the barrier width and thermal equilibrium is not disturbed by the measuring voltage. Here μ is the hole mobility, ψ is the electrical potential, and p_0 is the hole concentration at $\psi=0$. Now we suppose that $\sigma(x)$ is replaced with a step function (Fig. 3) where the σ_i's are constants. The lengths corresponding to σ_i are added and the sum is marked with l_i (Fig. 4). There have to be layers in every direction; thus the corresponding equivalent circuit would be a multiple connected network, but for computational reasons it has to be reduced to the form in Fig. 4.

In Fig. 4, σ_0 corresponds to the conductivity of selenium at $\psi=0$. The conductivity mechanism of selenium is active just in this area. Because l_0 is the greatest length and σ_0 the greatest conductivity, these have the strongest

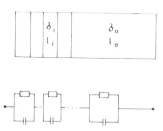

FIG. 4. The final model, according to which the results of the measurements were analyzed.

influence on the microwave conductivity, that is, the electrical properties of selenium are best measured at microwave frequencies according to this model. In this case one may regard the capacitances of the barrier layers as short circuits (Fig. 3).

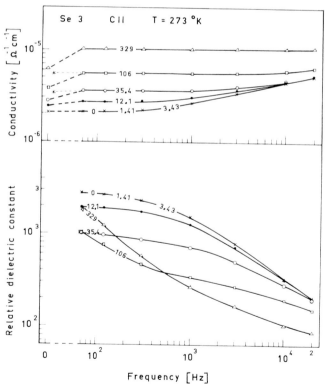

FIG. 5. Conductivity and relative dielectric constant at 273°K. The parameter of the curves is the mean d.c. electric field strength (V/cm). The a.c. field strength is about 0.5 V/cm. The curves can be fitted to the layer model, Fig. 7.

A layer model was constructed in accordance with Volger,[7] where the thickness and the conductivity of the layer were calculated as a function of the temperature and the bias potential. If the circuit in Fig. 4 is presented by means of the series parameters R_s and $1/C_s$, the dependence on l_i is of the first

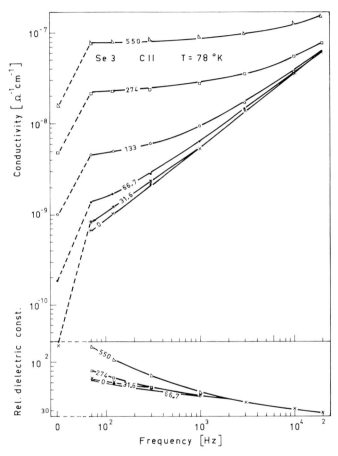

FIG. 6. Conductivity and relative dielectric constant at 78°K. The parameter of the curves is the mean d.c. electric field strength (V/cm). The a.c. field strength is about 0.5 V/cm. At "zero" field the conductivity is of the form which indicates a hopping process between impurity atoms.

order. Then it is possible to choose the σ_i's and solve the l_i's from the measurements. A layer with the conductivity σ_i has its greatest effect at the frequency $\omega_i = 1/\tau_i = \sigma_i/\epsilon_r\epsilon_0$. One has to fix σ_i so that it is equivalent to a certain ω_i-value within or near the measured frequency range. For ϵ_r the values in Fig. 9 were used, which were measured at 24 GHz. In accordance with this model one obtains a theoretical curve which differs only 5 per cent from the experimental curve (Figs. 5 and 7). More accurate discussions of the model and results are presented elsewhere.[13] Furthermore, one can prove that approximately 80 per cent of the entire crystal has a conductivity which corresponds with the value measured at 24 GHz.

At 78°K no solutions could be found for the different parameters, but on the other hand there is another peculiarity. The frequency dependence of the

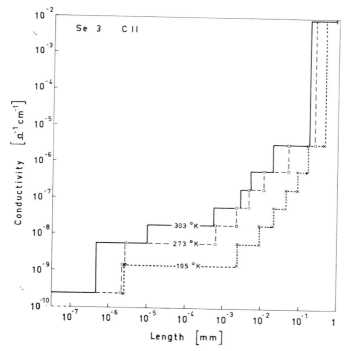

FIG. 7. Calculated conductivity (σ_i) of crystal regions versus length of region (l_i), when the layers are put in increasing conductivity order. Using parameters of the figure the maximum deviations are smaller than 5 per cent.

conductivity without biasing voltage was $\sigma \propto \omega^s$, where $s \simeq 0.8$, for all the crystals (Fig. 6). The same characteristics were found by Pollak and Geballe[8] for silicon at temperatures below 12°K. The phenomenon is supposed to arise from a hopping effect between the impurity atoms. The value $s = 0.8$ has been calculated theoretically.[8] The energy gap of selenium is 1.8 eV and the impurity density 30 ppm. Therefore, a hopping effect might occur in selenium at temperatures as high as $T = 78°K$.

In general, the d.c. as well as a.c. resistivity curves in semiconductors show the same tendency. This is valid especially for silicon and germanium, where the high-frequency resistivity ρ_f is obtained from the value of the d.c. resistivity ρ_0 by introducing the term $\tau/(1+j\omega\tau)$ instead of τ in the expression for ρ_0. The quantity τ represents the relaxation time and $\omega = 2\pi f$, where f is the measuring frequency. One can also point out a similarity between the curves for ρ_0 and ρ_f in tellurium.[9] It can then be presumed that the relatively flat part of the curve

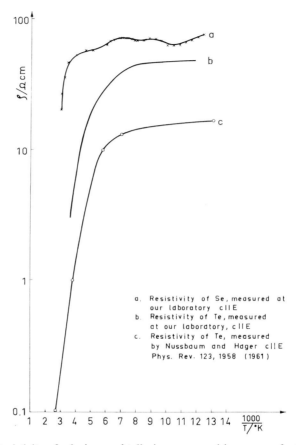

FIG. 8. Resistivity of selenium and tellurium measured by means of microwaves versus reciprocal temperature.

(c) at low temperatures (Fig. 8) represents an extrinsic conductivity caused by holes and that the steep positively inclined part at higher temperatures is caused by the intrinsic conductivity.[9] The curve (b) in Fig. 8 shows the resistivity of tellurium at 24 GHz with a measuring power 0.3 mW. The curves (b)

* The curve (a) is measured in nitrogen atmosphere. New measurements performed in vacuum have given results consistent with those of Lilja and Stubb, who have used argon atmosphere. The difference is assumed to depend on surface phenomena, because it depends on the nitrogen pressure.

and (c) are similar in shape. The resistivity curve for selenium measured at 24 GHz with a measuring power 0.3 mW is marked (a).

From various measurements we can establish some extraordinary processes going on in monocrystalline selenium at 200–230°K, i.e. within the region where the anomalies in the resistivity curve occur. Thus, when measuring current transients at constant voltage, one finds that the transients occur in the I–V characteristics above as well as below the mentioned temperature range, whereas no transients can be found within this temperature range.

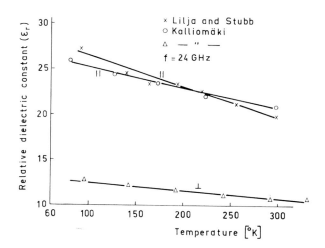

FIG. 9. Relative dielectric constant versus temperature measured by means of microwaves.

The dielectric constant in monocrystalline selenium was also measured at 24 GHz. The dielectric constant parallel to the c-axis was found to be $\epsilon_{r\|}=21$ and that perpendicular to the c-axis to be $\epsilon_{r\perp}=11$ at $T=300°$K. The values are reproduced graphically in Fig. 9.

The value $\epsilon_{r\|}=21$ has been criticized by Prosser,[11] who maintains that a wrong estimate of the electric field in the waveguide was made in our measurements. In order to refute this statement new measurements were made by means of a resonator method and the values obtained agreed with the previous results in which a bridge arrangement[12] was used.

REFERENCES

1. STUKE, J., *Phys. Status Solidi* **6**, 441 (1964).
2. HENKELS, H. V., *Phys. Rev.* **76**, 1737 (1949).
3. PLESSNER, K. W., *Proc. Phys. Soc.* **64B**, 681 (1951).

4. MORT, J., *Phys. Rev. Letters* **18**, 540 (1967).
5. JAUMANN, J. and NECKENBÜRGER, E., *Z. Physik*, **151**, 72 (1959).
6. MELL, H. and STUKE, J., *Phys. Letters* **20**, 222 (1966).
7. VOLGER, J., *Progress in Semiconductors* **4**, 207, Wiley, New York, 1960.
8. POLLAK, M. and GEBALLE, T. H., *Phys. Rev.* **122**, 1742 (1961).
9. NUSSBAUM, A. and HAGER, R. J., *Phys. Rev.* **123**, 1958 (1961).
10. GRAEFFE, R. and HELESKIVI, J., *Acta Polytechnica Scandinavica, Physics including Nucleonics*, Series No. 42, Helsinki, 1966.
11. PROSSER, V., *Recent Advances in Selenium Physics*, Pergamon, London, 1965.
12. LILJA, R. and STUBB, T., *Acta Polytechnica Scandinavica, Physics including Nucleonics*, Series No. 28, Helsinki, 1964.
13. SALO, T., "The Layer Model of Selenium Single Crystals", Thesis for the degree of Technical Licentiate (Electrical Engineering) (1967), Technical University, Electronics Laboratory, Helsinki, Finland.

DISCUSSION

EVERETT: Do you have an explanation for the decreasing dielectric constant with increasing temperature?

STUBB: The polarizability of permanent dipoles has the same qualitative temperature dependence.

AMITAY: Can you comment on the qualitatively opposite temperature dependence of the conductivities in your current work as compared to your previous paper with Lilja?

STUBB: If we had carried out the previous measurements also at higher temperatures, the curve published by Lilja and myself would probably have bent down in a similar way. Personally, I believe that there is some kind of dislocation effect that has an influence on the behavior of $\rho = \rho(T)$. (See also the footnote on page 342.)

ELECTRICAL CONDUCTIVITY OF SELENIUM AND SELENIUM-CONTAINING GLASSES AT TEMPERATURES UP TO 1000°C

R. W. HAISTY[†] and H. KREBS

Institut für anorganische Chemie der Universität Stuttgart (TH)

THE high-frequency electrical conductivity of melts of selenium, of the binary systems Se–Ge and Se–Sb, and of the ternary system Se–Sb–Ge has been measured in sealed quartz ampoules by an electrodeless "falling sample" method[1] in order to compare the conductivity of the melts of glass-forming compositions with that of compositions which solidify in crystalline form. It was also of interest to determine whether the semiconducting behavior of molten selenium, with an activation energy of 2.3 eV, which has previously been measured only up to 630°C[2] persists unchanged to higher temperatures.

The initial results at 14 mc/s showed an apparent maximum in conductivity around 600°C, with the conductivity then decreasing before turning up again around 800°C. Further work, however, has shown that this jog in the conductivity curve is caused by the transition, not taken into account in the original calibration curve,[1] from primarily inductive to primarily capacitive coupling as the sample conductivity decreases.[3] Correcting for this effect, it appears that the high-frequency conductivity continues to follow a temperature dependence corresponding to about 2.3 eV up to at least 900°C.

The Se–Sb–Ge melts having compositions corresponding to points inside the glass-forming region reported by Patterson and Brau[4] for this system were in all cases semiconducting up to 1000°C with relatively large energy gaps (*ca.* 2 eV). These compositions are indicated by the open circles in Fig. 1. Compositions just outside the easy glass-forming region formed melts having relatively high conductivity, with only a slight positive temperature dependence (half-filled circles in Fig. 1). These melts solidified partially as glasses, but only when quenched in water. With melts containing still more germanium or antimony, distinct metallic-like conductivity was found.

A pronounced tendency to form glasses even upon slow cooling is only possible if the bonding arrangement of the atoms in the melt coincides with that of the cooled glass. We are dealing with more or less highly polymeric

[†] On leave of absence from Texas Instruments, Inc., Dallas, Texas.

networks up to several hundred degrees above the softening point, of which the semiconducting behavior is a consequence. Other conduction mechanisms, such as ionic conductivity, are, in general, not found in the semiconducting chalcogenide glasses. The mobility of the atoms above the softening point is a result of the large number of bridging atoms (selenium here) present.[5] If the three-dimensional networks become well developed by the addition of sufficient quantities of 3- and 4-valent atoms such as antimony and germanium, the

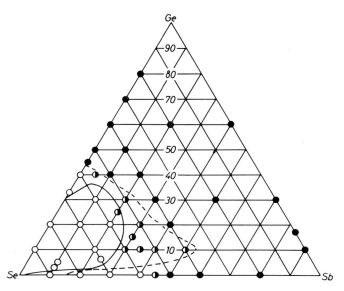

FIG. 1. Distribution of semiconducting and metallic-like melts in the Se–Sb–Ge system. Open circles indicate typical semiconductor behavior, half-filled circles represent samples having relatively high conductivity, but with a positive temperature dependence, and filled circles show metallic conductivity. The smaller glass-forming region shown is that reported by Patterson and Brau[4] for air quenching. The larger region includes those samples found in this work to solidify in primarily glassy form upon water quenching.

mobility of the atoms becomes so restricted that melting is only possible by a change in the bonding system. The bonding in the network which takes place via (sp^x)-hybrid states must, upon melting, go over into bonding which consists primarily of valence electrons in p-states. The coordination number of the elements then approaches the value 6, and there result melts with low viscosities and metallic electrical conductivities. This has recently been checked with melts of pure germanium.[6]

Semiconducting behavior in melts can have several different causes, and not all of these are connected with the conditions for easy glass formation.[7] In the binary Sb–Se system, even small amounts of antimony appear to induce crystallization, similar to the behavior of traces of potassium in selenium.[5] These melts have much higher conductivities than those of pure selenium, but are still not metallic in nature. Thus, the absence of metallic conductivity in the melt is not a sufficient condition for easy glass formation, but it appears from the present results that it is a necessary one.

The relatively large high frequency conductivity of the samples bordering the glass region, which are still semiconducting, is probably due to the existence of a sub-microscopically divided second phase. Micelles of one phase are suspended in a matrix of the other phase.

REFERENCES

1. HAISTY, R. W., *Rev. Sci. Inst.* **88**, 262 (1967).
2. JOFFE, A. F. and REGEL, A. R., *Prog. in Semiconductors* **4**, 264 (1960).
3. HAISTY, R. W., *Rev. Sci. Inst.* **39**, 778 (1968).
4. PATTERSON, R. J. and BRAU, M. J., presented at The Electrochemical Society Meeting, May 1966, Cleveland, Ohio.
5. KREBS, H., *Angew. Chem. Int. Ed.* **5**, 544 (1966).
6. KREBS, H., LAZAREV, V. B. and WINKLER, L., *Z. Anorg. Allgem. Chem.* **352**, 277 (1967).
7. HAISTY, R. W. and KREBS, H., to be published in *J. Non-crystalline Solids*.

DISCUSSION

STUKE: What is the difference between red amorphous and vitreous selenium?

KREBS: One obtains red amorphous selenium by the oxidation of H_2Se or by the reduction of selenium compounds such as SeO_2 in non-alkaline solutions at room temperature. It consists of small ring molecules, predominantly Se_8, since it is completely soluble in CS_2 and crystallizes out as α- or β-monoclinic red selenium, which consists of Se_8 rings (see Gmelin, *Handbook of Inorganic Chemistry*, 8th edition, System No. 10, Selenium, Part A, pp. 100 and 257).

In glassy selenium we have, in addition to the Se_8 rings, large rings. The latter are entangled with each other, so that a solid glassy substance results, as opposed to the more loosely packed red amorphous selenium, which is obtained as a powder.

ELECTRICAL BEHAVIOR OF THE CONTACT BETWEEN CADMIUM AND A SINGLE CRYSTAL SELENIUM FILM

C. H. CHAMPNESS,[†] C. H. GRIFFITHS[‡] and H. SANG[§]

Noranda Research Centre, Pointe Claire, Quebec, Canada

INTRODUCTION

Despite development over some 40 years, the selenium rectifier is still inferior to silicon and germanium diodes in all respects except two. These are firstly, that the selenium rectifier is relatively simple to manufacture and therefore can be made cheaply; and secondly, that it possesses the valuable property of being able to recover from high-voltage surges, allowing reliable application in series and parallel connections. The device uses a polycrystalline film of trigonal selenium. Since most of the progress in silicon and germanium devices has been accomplished through the controlled use of these elements in single crystal form, it was wondered if a parallel development of selenium devices could be obtained through the use of single crystals of trigonal selenium.

Early attempts to grow large single crystals were unsuccessful, but recently Harrison and Tiller[1] were able to grow large well-oriented ingots from the melt under a pressure of about 5 kilobars. Single crystals were also grown by Keezer using a travelling solvent method and by Keezer and Moody[2] using the Czochralski method in which impurities were introduced into the selenium to shorten the chains and thus facilitate their alignment. However, such bulk crystals do not readily lend themselves to the rectifier application where a thin film of selenium is needed. As described elsewhere,[3, 4] a method of preparing single crystal trigonal selenium films in the ($10\bar{1}0$) and (0001) planes was developed in this laboratory, essentially by epitaxial deposition on single crystal tellurium. Simple rectifier units were then prepared by deposition of cadmium on top of the selenium film.

Figure 1 shows the form of the single crystal rectifier units compared with that of a commercial polycrystalline rectifier. In the latter case the selenium

---

[†] Present address: Department of Electrical Engineering, McGill University, Montreal, Quebec, Canada.
[‡] Present address: Research and Engineering Center, Xerox Corporation, Webster, N.Y.
[§] Present address: Department of Metallurgy and Materials Science, University of Toronto, Toronto 5, Ontario, Canada.

film is normally about 50 microns in thickness and consists of randomly oriented grains of the trigonal form which have crystallized through spherulitic growth from the amorphous form. The base contact is steel or aluminum and the counter electrode is usually an alloy of cadmium and tin. Many variations exist for the introduction of additives at the base and counter electrode contacts to the selenium. In the single crystal units, the base was the tellurium substrate on which the film was grown, ensuring a low-resistance ohmic contact at this junction, while the counter electrode was of pure cadmium. By the use of single crystal films it was hoped that lower forward resistances could be obtained resulting from their perfect orientation and smaller thickness.

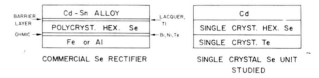

FIG. 1. Cross-sections showing the layer structure of a commercial selenium rectifier and the type of single crystal rectifier investigated in this work.

It is now generally accepted that rectification in a selenium rectifier arises not from a metal-semiconductor junction but from an n-type CdSe, p-type Se heterojunction.[5] The CdSe is formed by reaction between the cadmium and the selenium although some manufacturers actually deposit a layer of CdSe (and also CdS) on the selenium. The results of measurements on single crystal units in this paper are compared with what would be expected for such a junction.

THEORY

Before describing the experimental work it is convenient to write down the appropriate relations for the current density, incremental resistance and capacitance of a rectifier.

Current–Voltage Relation

In a rectifier the relation between current density (j) and applied voltage (V), both taken as positive in the forward direction, can generally be expressed in the form:

$$j = j_0 \left\{ \exp\left[\frac{e(V-jR)}{akT}\right] - 1 \right\}, \tag{1}$$

where R is the bulk specific area resistance of the material constituting the rectifier, T is the absolute temperature, e is the electronic charge and k is Boltzmann's constant. The quantity a is a coefficient equal to unity for a

Schottky metal-semiconductor contact or a *p–n* junction with negligible recombination in the junction region.

Symmetrical heterojunction. Dolega[6] has shown for a perfectly symmetrical heterojunction that $a = 2$ and that j_0 is given by

$$j_0 = \sigma \left(\frac{\pi e n}{2\kappa}\right)^{\frac{1}{2}} (V_D - V)^{\frac{1}{2}} \exp\left(-\frac{eV_D}{2kT}\right), \tag{2}$$

where σ is the conductivity, n is the density of ionized charge carriers in the depletion layer and κ is the relative dielectric constant for both the *n*-type and *p*-type material assumed identical in these respects. A reverse saturation current, such as exists in a *p–n* junction, does not arise here because the high recombination rate in a heterojunction prevents the injection of minority carriers from one semiconductor into the other.

Reverse Incremental Resistance for a Symmetrical Heterojunction

The reverse incremental resistance R_{inc} ($= dV_R/dj_R$, where $j_R = -j$ and $V_R = -V$) for a symmetrical heterojunction is obtained from equations (1) and (2) by differentiation neglecting the term jR, which is small. The result when $V_R \gg 2kT/e$ is

$$\frac{R_{\text{inc}}}{R_0} = \frac{e}{kT}[V_D(V_D + V_R)]^{\frac{1}{2}}, \tag{3}$$

where

$$R_0 = \frac{2kT}{\sigma e}\left(\frac{2\kappa}{\pi e n V_D}\right)^{\frac{1}{2}} \exp\left(\frac{eV_D}{2kT}\right). \tag{4}$$

Thus for $V_R \gg V_D$ the incremental resistance in a symmetrical heterojunction varies approximately as the square root of the voltage. In an unsymmetrical heterojunction, however, Dolega[6] has shown that R_{inc} increases to a maximum value.

Capacitance–Voltage Relation

For a heterojunction containing a uniform distribution of n_A acceptors in the *p*-type material and n_D donors in the *n*-type material, Dolega[6,7] has shown that the relation between the parallel capacitance per unit area C and the applied reverse voltage, V_R ($= -V$) has the form

$$\frac{1}{C^2} = \frac{8\pi}{e}\left(\frac{1}{\kappa_n n_D} + \frac{1}{\kappa_p n_A}\right)(V_D + V_R). \tag{5}$$

For a symmetrical heterojunction $\kappa_n = \kappa_p = \kappa$ and $n_D = n_A = n$ so that

$$\frac{1}{C^2} = \frac{16\pi}{e\kappa n}(V_D + V_R). \tag{6}$$

FABRICATION OF SINGLE CRYSTAL SELENIUM RECTIFIER UNITS

A number of single crystal rectifying tellurium–selenium–cadmium structures (Fig. 1) were prepared in the laboratory and Table 1 shows some details of the procedure for making typical units. The method of fabrication was basically as follows:

Tellurium Substrate

A slice of about 3 mm in thickness was cut in a ($10\bar{1}0$) or (0001) plane from a bulk crystal of tellurium grown either by horizontal zone melting or by the Czochralski method. It was then mounted in plaster of paris and polished mechanically using, in the final stage, stannous oxide powder with water. After polishing, the plaster was chipped away and the tellurium slice was cleaned and then mounted on a substrate heater in a bell-jar vacuum system. It was then annealed to remove surface damage by heating in a vacuum of 10^{-5} torr. The treatment was 2 hr at 200°C for the zone-melted tellurium but a lower temperature was used for the Czochralski-grown tellurium.

Selenium Deposition

The technique for the epitaxial growth of the selenium film has been described in detail.[4] Briefly, in this case the method was to evaporate the selenium from a quartz boat onto the tellurium substrate maintained at a temperature near 100°C and at a rate of between 0.2 and 0.5 micron/min. The time of deposition was between about 5 and 30 min depending on the film thickness required. For the first units prepared, undoped selenium was used but in later experiments selenium doped under vacuum with $SeCl_4$ was employed.† Cracking of the films sometimes occurred but this was found to be minimized by slow cooling of the film over several hours following evaporation.

Extensive electron diffraction and X-ray diffraction evidence was obtained to prove that the selenium film was in fact monocrystalline.[4] Figure 2 shows a Laue back reflection X-ray photograph taken on a (0001) selenium film attached to a tellurium substrate from which a rectifier unit (No. 24 (c)) was made. The same array of spots showing three-fold symmetry was obtained at a number of positions across the film indicating no change in the orientation. The fact that the pattern for this thickness was from the selenium rather than the underlying tellurium was proved from X-ray experiments on selenium films

---

† This was done using a quartz vessel with three arms, of which one contained selenium, the second $SeCl_4$ and the third was empty. The selenium containing the $SeCl_4$ was introduced into the empty tube by rotating and heating the vessel about a ground-glass connection to the evacuating system in two movements.

FIG. 2. Laue back reflection X-ray photograph taken at several points on a (0001) selenium film on a tellurium substrate forming part of a rectifier unit.

TABLE 1. *Details of Preparation of Typical Selenium Single Crystal Rectifier Units*

Unit no.	Selenium used	Te single crystal used as base	Selenium deposition				Forming		Remarks
			Substrate temp. (°C)	Time (min)	Se film thickness (micron)		Peak reverse voltage	Time (hr)	
12	Undoped commercial "high purity"	Polished (10$\bar{1}$0) zone-melt grown	94	8	8.4		90	24	High reverse "breakdown" with high forward resistance
17	200 ppm Cl-doped under vacuum	"	93.5	30	—		50	91	—
20 (a) (b)	"	"	94	5 11	5.6 10.6		50 40	96 109	Unit evaporated with two thicknesses of selenium and two separate strips of cadmium
26	3000 ppm Cl-doped	"	93.5	20	7.7		40	97	Selenium with strong Cl doping used
30	200 ppm Cl-doped	Polished (0001) Czochralski grown	103	9	—		65	92	—

Se deposition: Substrate to source distance, ~ 12 cm. Boat charge, 2 to 5 g.

Cd deposition: Substrate to source distance, ~ 10 cm. Substrate at room temperature. Boat charge, ~ 5 g.

completely removed from the substrate.[4] The thickness of the selenium films was determined *in situ* by observation of the interference reflection maxima using a model 13U Perkin–Elmer spectrophotometer.

Cadmium Deposition

Using pure cadmium, this metal was evaporated under vacuum onto the selenium film at room temperature. The area of metal deposited on the selenium was confined by a hole in stainless-steel foil but in earlier experiments this masking was done using collodion directly on the selenium and tellurium substrate.

FORMING

Following the fabrication of a unit, short circuits were often found between the cadmium and the tellurium. By passing a pulse of current through the unit in the forward direction these could sometimes be "burned out". The process of "forming" the rectifier was then carried out in the same way as in a polycrystalline rectifier by passing rectified alternating current through the unit in the reverse direction for a number of hours. As the reverse characteristic improved with time, the applied voltage was increased. This was continued until there was no further change. During the forming, short circuits were frequently found to occur in spots visible to the eye. However, it was found that these could be removed by scraping away or sectioning off the cadmium in the regions concerned. Following this, the rectification action was restored but with a smaller effective area. Thus after forming in such cases the total forward resistance would be higher than at the start. Figure 3 shows oscilloscope current–voltage characteristics taken on a unit before and after forming. The increase in the reverse voltage which could be applied without appreciable rise in current is clearly seen.

The electrical contacts consisted of a spring-loaded carbon rod on the counter electrode and a pressure or soldered contact at the tellurium base to a large plate of brass.

DYNAMIC CURRENT–VOLTAGE CHARACTERISTICS

The most noticeable feature of the current–voltage characteristic of a single crystal unit compared with that of a commercial polycrystalline rectifier is the curvature in the forward direction. This is clearly shown in Fig. 4 where it is seen that beyond the "knee" at about 0.5 volt the forward characteristic of the commercial rectifier is quite straight, whereas that for the single crystal unit is strongly curved. This effect is further studied in Fig. 5 where the forward dynamic characteristics above about 1 volt are plotted on log-log scales for a number of single crystal units. It is apparent in this region that the current

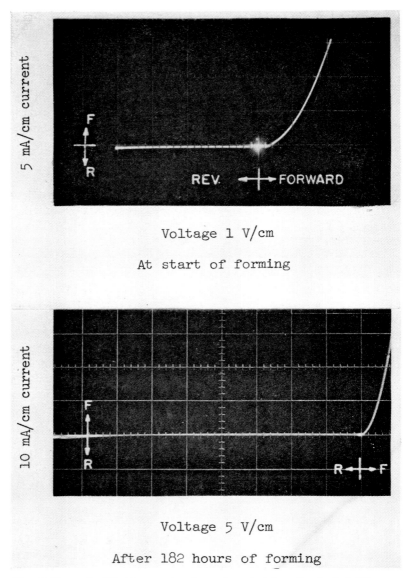

FIG. 3. Dynamic (60 c/s) current–voltage characteristics of a typical 200 ppm chlorine-doped (0001) single crystal selenium rectifier unit.

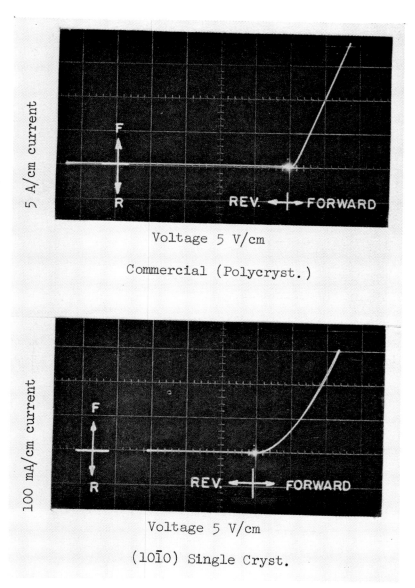

FIG. 4. Comparison of the shapes of dynamic (60 c/s) current–voltage characteristics of a typical commercial selenium rectifier and a single crystal unit.

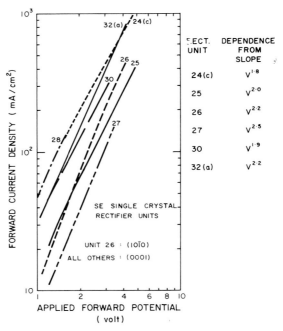

Fig. 5. Log-log plot of current density against applied forward potential above 1 volt for six single crystal rectifier units, measured dynamically at 60 c/s from oscilloscope photographs.

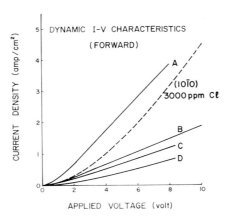

Fig. 6. Plot of current density, measured dynamically, against applied forward voltage for four commercial selenium rectifiers A, B, C and D, compared with that for a single crystal ($10\bar{1}0$) rectifier unit.

density has a nearly quadratic dependence on voltage for films in both the (10$\bar{1}$0) and (0001) planes and suggests space charge limited current behavior.

Figure 6 shows dynamic current density–voltage characteristics on linear scales of several commercial selenium rectifiers in the forward direction compared with that of a (10$\bar{1}$0) single crystal unit. It is seen that with one exception the latter showed a higher current density in the range 2 to 10 volts than the commercial units studied.

STATIC CURRENT–VOLTAGE CHARACTERISTICS

Following the dynamic current–voltage characteristics, static measurements were made on the units using direct current. Figure 7 shows static characteristics measured on two units prepared with different film thicknesses on the same tellurium base. While the reverse characteristics are essentially the same, the 5.6 micron unit (20 (a)) shows a much greater current density in the forward direction. This indicates clearly that it is the thickness of the selenium layer rather than that of the CdSe layer which controls the forward resistance.

Figure 8 shows a theoretical plot of ln j versus V in the forward direction from equation (1) assuming a constant value for j_0 and two values for R. The two curves are seen to have three parts; an ohmic part at low voltage with the slope resistance R_0 ($=akT/j_{00}$, where j_{00} is the value of j_0 at $V=0$); another ohmic part at higher voltages where the resistance is R; and an exponential

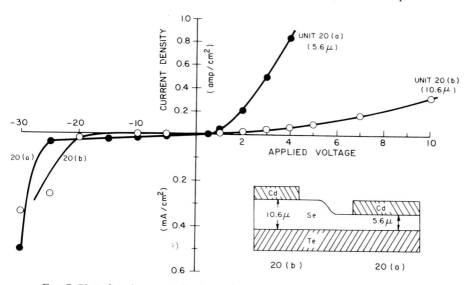

FIG. 7. Plot of static current density–voltage characteristics for two single crystal (10$\bar{1}$0) units with different film thicknesses mounted on the same tellurium substrate.

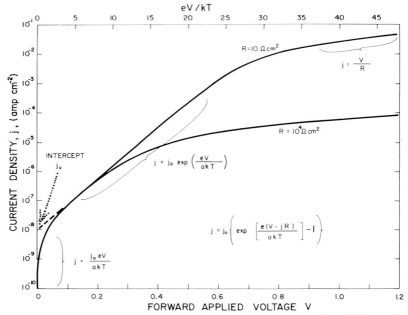

FIG. 8. Plot on log-linear scales of the variation of current density with forward applied voltage calculated from the rectifier equation assuming $j_0 = 10^{-8}$ amp/cm^2 and two values of R.

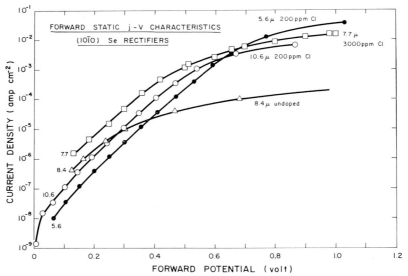

FIG. 9. Plot on log-linear scales of the variation of static current density with applied forward voltage for four $(10\bar{1}0)$ single crystal rectifier units.

part (linear on this plot) between these two extremes. It is to be noted that the slope of the exponential region depends on the value for the a-coefficient.

The experimental plot of $\ln j$ versus V shown in Fig. 9 for four $(10\bar{1}0)$ single crystal units indicates clearly these three regions. For one of the curves the exponential region is seen to extend over some five decades of current density. As expected from Fig. 8, the curves where the length of this region is shorter show a lower current density above 0.8 volt arising from a higher resistance R. Near 0.8 volt the curve for the unit with selenium containing 200 ppm of

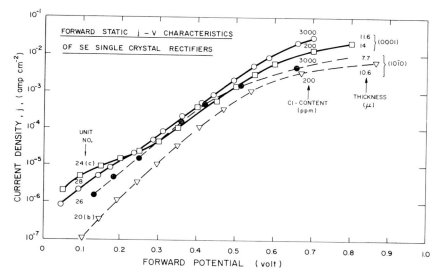

FIG. 10. Log-linear plot of static current density against forward applied potential for chlorine-doped single crystal rectifier units showing the difference between the $(10\bar{1}0)$ and (0001) orientations.

chlorine and film thickness of 10.6 microns lies about two orders of magnitude in current density above that for the unit with zero chlorine content. However, the increase is much less marked for the further addition of chlorine from 200 to 3000 ppm. It is to be noted that above about 0.8 volt the current densities for the chlorine-doped units increase with decreasing film thickness. Figure 10 shows a comparison of $\ln j$–V curves for $(10\bar{1}0)$ (broken lines) and (0001) units (solid lines). It is clear that at higher voltage the latter have almost an order of magnitude higher current density, due to the greater conductivity parallel to the trigonal axis than at right angles to it. The slopes, measured from the exponential regions of Figs. 9 and 10 and other plots not shown, give a-values near 2 (see Table 2, column 14). By extrapolation from the exponential regions to the $V=0$ axis, values of j_{00} were obtained. Some of these are listed in Table 2, column 6, and are seen to be of the order of 10^{-7} to 10^{-8} amp/cm².

TABLE 2. Measured Parameters on Selenium Single Crystal Rectifier Units

Unit no.	Cl content (ppm)	Se film orient.	Active area (cm^2)	d_{Se} (μ)	j_0 intercept at $V=0$ (μA/cm^2)	Rev. b.d. pot. (volt)	V_D (volt)	Thick S.C.L. at 0 volt (μ)	n_A (10^{16} cm^{-3})	Spec. area forw. slope res. (Ω cm^2)	Forward res. (10^3 Ω cm)	Eff. mobility (cm^2 V^{-1} s^{-1})	$j=j_0$ exp (eV/akT) a
12	0	(10$\bar{1}$0)	1.0	8.4	0.02	70	1.5	0.28	1.2	670	800	6.5×10^{-4}	1.5
15	100	,,	0.5	—	0.01	45	0.5	0.1	3.3	15	—	—	1.7
17	200	,,	0.5	—	0.01	40	0.7	0.11	3.4	20	—	—	2.0
20 (a)	200	,,	0.08	5.6	0.004	40	0.8	0.12	3.5	11	19	9×10^{-3}	1.8
20 (b)	200	,,	0.09	10.6	0.016	40	0.5	0.09	4.0	23	22	7×10^{-3}	1.8
26	3000	,,	0.065	7.7	0.1	30	0.35	0.08	4.0	4.5	6	2.6×10^{-2}	2.0
24 (c)	200	(0001)	0.07	14	0.1	40	0.8	0.13	3.2	2.2	1.6	1.2×10^{-1}	—

CAPACITANCE AND INCREMENTAL RESISTANCE MEASUREMENTS

Measurements of parallel capacitance C and reverse incremental resistance R_{inc} were made as a function of reverse voltage using a General Radio model 1608A impedance bridge and associated apparatus (Fig. 11). During the measurements it was found that the capacitance readings increased with time as shown in Fig. 12. The same effect was reported in GaAs Schottky diodes by Furakawa and Ishibashi,[8] who found that their plots of $(A/C)^2$ against voltage were only linear when the C-values were extrapolated to zero time. Accordingly the same extrapolation procedure was adopted in the present case and Fig. 13 shows typical results for three units where linearity is apparent from 0 to 2 volts.

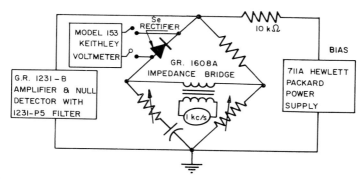

FIG. 11. Bridge circuit used for the measurements of parallel capacitance and incremental resistance on the rectifiers.

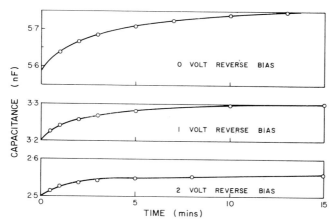

FIG. 12. Plot showing typical examples of the increase of capacitance with measurement time for the single crystal (10$\bar{1}$0) selenium rectifier unit 20 (b).

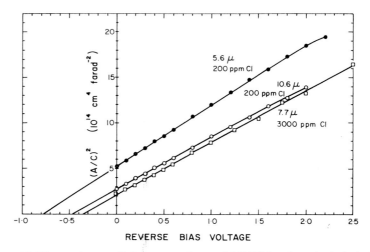

Fig. 13. Plot against applied reverse voltage of $(A/C)^2$ for three typical single crystal rectifier units, where A is the rectifier area and C is the parallel capacitance extrapolated to zero measuring time.

Values of V_D from the intercepts on the voltage axis are given in Table 2, column 8 and are seen to lie between 0.35 and 0.8 volt for the units with chlorine-doped selenium. Values of carrier concentration n_A were deduced from the slopes using equation (5) assuming $n_D \gg n_A$ and these are given in Table 2, column 10. It is noted that these values are all of the order of 3×10^{16} cm^{-3} irrespective of the nominal amount of chlorine in the selenium.

Figure 14 shows a plot against reverse voltage of the ratio of incremental resistance to that at zero potential ($= R_{\text{inc}}/R_0$). The rate of increase is smaller than predicted by equation (3) and a maximum is clearly apparent near 2.5 volts for one of the units.

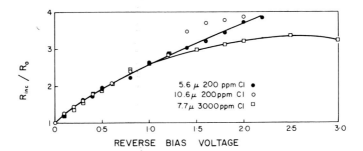

Fig. 14. Plot against applied reverse voltage of the ratio of the incremental resistance (R_{inc}) to that at zero voltage (R_0) for three ($10\bar{1}0$) single crystal rectifier units.

MEASUREMENTS ON Te–Se–Te STRUCTURES

Two Te–Se–Te structures were prepared by epitaxial deposition of selenium on single crystal tellurium followed by deposition of tellurium on the selenium. One unit used undoped selenium with the film oriented in the ($10\bar{1}0$) plane while the other used selenium doped with 200 ppm chlorine oriented in the (0001) plane. Static current–voltage characteristics taken on these units between

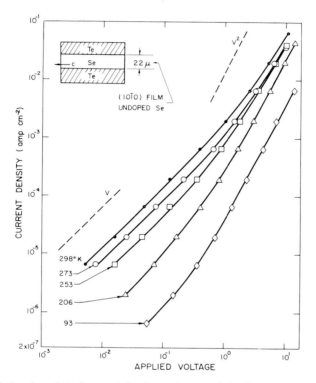

FIG. 15. Log-log plot of current density against applied voltage at different temperatures for a Te–Se–Te structure using undoped selenium oriented in the ($10\bar{1}0$) plane.

room and liquid nitrogen temperature are shown in Figs. 15 and 16. Below about 0.1 volt the dependence of current density on voltage is seen to be linear but at higher voltages a quadratic dependence is apparent. The transition from one régime to the other occurs at lower voltages the lower the temperature of the sample. These results indicate in a single crystal rectifier firstly that the Se–Te contact is non-rectifying and secondly that the V^2 dependence at higher forward voltage is not associated with the presence of the cadmium or the rectification process but with the properties of the selenium film itself.

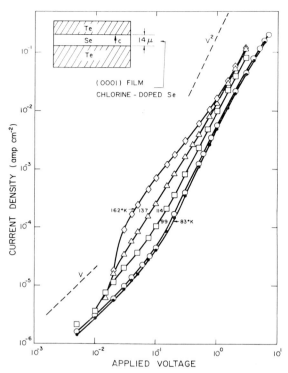

FIG. 16. Log-log plot of current density against applied voltage at different temperatures for a Te–Se–Te structure using 200 ppm chlorine-doped selenium oriented in the (0001) plane.

For space charge limited currents (S.C.L.C.) in the simplest case, the current density is given by

$$j = \frac{9\kappa\mu V^2}{8d^3}, \tag{7}$$

where μ is the carrier mobility and d is the thickness of the sample. Application of this equation to the results in Figs. 15 and 16 gives mobility values between 10 and 100 cm^2 volt^{-1} sec^{-1}. Similar values are also obtained from the forward characteristics of the rectifier units in Fig. 5. Such values are thus two or more orders of magnitude greater than those deduced from the capacitance data in Section 7.

DISCUSSION

The results obtained on a number of single crystal rectifier units are summarized in Table 2. The fact that the a-values were observed to be about 2 is consistent with the idea that the rectification occurs at a nearly symmetrical heterojunction rather than at a Schottky metal-semiconductor junction. However, the fact that R_{inc}/R_0 does not increase montonically with reverse voltage according to equation (3) but shows a maximum, indicates that the heterojunction cannot be perfectly symmetrical. This is not surprising since the values of n_D, μ_n and κ_n for n-type CdSe are likely to be different from n_A, μ_p and κ_p for p-type Se. The marked decrease in forward resistance with decrease of film thickness indicates that the CdSe layer is unimportant for this direction and must therefore be very thin.

Using the values $V_D = 0.5$ volt, $n = 4 \times 10^{16}$ cm^{-3} and $\kappa = 10$ in equation (2) gives a value for j_0 of 3.5×10^{-4} amp/cm² at $V = 0$. This, however, is more than three orders of magnitude greater than the j_0 intercept values given in column 6 of Table 2, a fact not yet understood.

The carrier concentrations as deduced from the capacitance plots (column 10) are seen to be essentially independent of the nominal chlorine content of the film as well as of the thickness and orientation. This could happen if the number corresponded to n_D rather than n_A but this would require $n_A \gg n_D$ which seems to be unlikely. Nevertheless, the chlorine does decrease the forward resistivity (column 12) by increasing the apparent mobility (column 13). It should be noted at this point that since the values in column 11 were obtained from the slope of the characteristics in the forward direction where a quadratic dependence on voltage holds, these resistance values and the resistivity values in the next column should be regarded as underestimates. By the same token, the mobility values in column 13 must be overestimates. The discrepancy between these low mobility values and the much higher ones between 10 and 100 cm² volt^{-1} sec^{-1} obtained by applying the simple S.C.L.C. equation is another case of the same problem that has occurred in other studies on trigonal selenium where one type of measurement (e.g. Hall) yields a mobility of about 0.1 cm² volt^{-1} sec^{-1} while another (magnetoresistance) yields a value some two orders of magnitude greater.

The fact that the quadratic dependence of current density on voltage occurred in both the Te–Se–Te and Te–Se–Cd structures indicates that the effect has nothing to do with the cadmium or the rectification process but is a property of the selenium film itself. Further, the ohmic dependence in the Te–Se–Te structures at low voltage confirms that the Te–Se base contact in a rectifier is also ohmic.

The resistivities in the single crystal films appear to be higher than in corresponding polycrystalline material. This may arise from conduction along grain boundaries in the latter case. Despite such an effect, it would seem that higher forward current densities may be obtained in single crystal rectifiers by making the film sufficiently thin. This arises from the fact that the current density for S.C.L.C. behavior varies as the inverse cube or inverse square[9] of the thickness. The ultimate limit on thickness is probably set by structural and preparational conditions but from a theoretical point of view, the minimum would appear to be the thickness of the CdSe layer (~ 0.2 micron) or the thickness of the depletion layer in the selenium for a large reverse voltage, estimated to be of the order of 1 micron. Thus if a (0001) rectifier unit free of short circuits could be fabricated with a film thickness of say 2 microns, it should exhibit a higher forward current density than is possible in a polycrystalline selenium rectifier. Of course, for practical application such performance must be accompanied by a retention of the valuable self-healing property which has not been demonstrated so far in single crystal units.

SUMMARY

Single crystal rectifiers have been made with the selenium film in both the prism and basal planes.[10] These have shown reverse characteristics comparable to and forward characteristics generally better than those of commercial polycrystalline rectifiers. The forward current density was found to vary quadratically with voltage, to increase with decrease of film thickness and to be greater for the (0001) than the ($10\bar{1}0$) orientation.

ACKNOWLEDGMENT

The authors would like to acknowledge the support of Canadian Copper Refiners Limited and the Directorate of Industrial Research of the Defence Research Board of Canada.

REFERENCES

1. HARRISON, D. E. and TILLER, W. A., *J. Appl. Phys.* **36**, 1680 (1965).
2. KEEZER, R. C. and MOODY, J. W., *Appl. Phys. Letters* **8**, 233 (1966).
3. GRIFFITHS, C. H. and SANG, H., *Appl. Phys. Letters* **11**, 118 (1967).
4. GRIFFITHS, C. H. and SANG, H., *The Physics of Selenium and Tellurium*, Pergamon, New York, 1969, p. 135.
5. HOFFMAN, A. and ROSE, F., *Z. Physik* **136**, 152 (1953).
6. DOLEGA, U., *Z. Naturforsch* **18a**, 653 (1963).
7. DOLEGA, U., *Z. Physik* **167**, 46 (1962).
8. FURAKAWA, Y. and ISHIBASHI, Y., *Japan J. Appl. Phys.* **5**, 837 (1966).
9. GRAEFFE, R. and HELESKIVI, J., *Acta Polytechnica Scandinavica, Physics including Nucleonics*, Series No. 42, Helsinki (1966).
10. GRIFFITHS, C. H., CHAMPNESS, C. H. and SANG, H., Canadian Patent No. 799,644, November, 19, 1968.

DISCUSSION

M. A. NICOLET (California Institute of Technology): If the model of S.C.L.C. applies and if your values of hole mobility are, somewhat generously, taken to agree with the accepted value for free holes, this means that trapping is absent in your samples. This would mean that trigonal Se is the third crystal in which pure S.C.L.C. is observed under steady state conditions, Ge and Si being the other two. It is also significant that mobility values obtained from

S.C.L.C. are largely independent of uncertainties in the doping properties of the bulk. It is not simple to firmly establish the presence of S.C.L.C. though. Two questions: How do you apply the contacts of Cd and Te after the crystal is grown? Have you checked the electrical symmetry of your Te–Se–Te devices?

CHAMPNESS: As described in the paper, the selenium film was deposited on bulk-grown single crystal tellurium and the cadmium or tellurium films were deposited on top of this by evaporation under a vacuum of 10^{-5} torr.

Answering your second question, no, we did not check in detail the symmetry of the current–voltage-dependence of the Te–Se–Te structures, but we did observe the dependence in both current directions.

S. WALLDÉN (European Selenium–Tellurium Committee and Bolidens Gruvaktiebolag): Do you know anything as to the position of the chlorine in the molecular structure?

CHAMPNESS: It would seem that some chlorine enters the selenium single crystal because of the observed increase in the forward current density. However, large quantities do not seem to produce a proportionate increase. I would not like to comment on where the chlorine is actually situated in the solid. Perhaps Dr. Keezer would like to comment on this.

KEEZER (Xerox Corporation): We do not have any evidence that impurities except tellurium and selenium are incorporated in the selenium single crystal lattice.

WALLDÉN: Have you observed any diffusion between the selenium and tellurium?

CHAMPNESS: We have not looked into this matter, but electron diffraction patterns from selenium films stripped from the tellurium substrate have shown no evidence of the presence of tellurium.

WALLDÉN: Have you any alternative method to suggest for the introduction of halogens such as diffusion or neutron treatment?

CHAMPNESS: We have only tried doping by the introduction of $SeCl_4$ into the selenium before evaporation. It is quite possible that exposure to halogen vapor could be effective.

H. GOBRECHT (Technischen Universität, Berlin): The epitaxial layers are thinner than the polycrystalline ones. Therefore my question is, will the lifetime of the rectifiers not decrease because of the thinner layers? The various manufacturers lay emphasis more or less on long-life of the rectifiers.

CHAMPNESS: We have not yet studied the aging properties of the structures. So far as I know, aging results from changes at the interfaces between the selenium and the base and counter electrodes. Thus the thickness of the selenium would appear not to be directly involved. We believe that no aging should arise from the base contact in our structure since this involves the excellent low resistance Se–Te bonding.

GOBRECHT: Is not the current density in Fig. 6 obviously a function of the layer thickness?

CHAMPNESS: Yes, the current density in the forward direction increases markedly with decrease of thickness of the selenium, but we do not have enough data to determine the exact form of the dependence at the present time.

H. P. D. LANYON (Worcester Polytechnic Institute): If one has a blocking contact, the first V_D of applied potential goes to fill the exhaustion layer. Only then does one start to inject excess charge. Consequently, j is proportional to $(V-V_D)^2$ not V^2.

CHAMPNESS: Yes, this would seem to be correct. Since we did not have V_D values for all the units, $\ln j$ was plotted simply against $\ln V$. In the cases where V_D was measured, a plot against $\ln (V-V_D)$ showed also an approach to a quadratic dependence at higher voltages.

LANYON: Capacitance measures total charge removed from surface region. This is both free and trapped, not only free. Consequently, one would expect the effective mobility to be much lower for this case than for the forward sense.

CHAMPNESS: As pointed out in the paper, the mobilities deduced from the carrier concentration of about 3×10^{16} cm^{-3} obtained from the capacitance data are overestimates and thus are exceedingly small. Higher mobility values would be obtained by assuming a smaller active concentration of free carriers for forward current flow, as you suggest.

LANYON: In deriving the forward j–V characteristic, you assumed the symmetric case of $n_D = n_A$, $\kappa_A = \kappa_p$ and $\mu_n = \mu_p$, but in deriving the capacitance you assumed the completely non-symmetric case $\kappa_n n_D \gg \kappa_p n_A$. This does not seem to be self-consistent.

CHAMPNESS: Yes, this was only done for convenience. The true situation is probably that while the junction is a heterojunction it is not perfectly symmetrical. The symmetrical case just produces a simpler formula and indicates an a-value of 2. For capacitance we cannot of course consistently assume $\kappa_n n_D \gg \kappa_p n_A$, but even if we assume $\kappa_n n_D = \kappa_p n_A$, we still get an apparent carrier concentration, free and trapped as you have pointed out, of about 10^{16} cm^{-3}.

J. MORT (Xerox Corporation): Did you actually see transition to trap-filled limit in your space-charge limited current? In other words, did you reach the Child's law region? This is the only region in which you can reliably estimate mobility values.

CHAMPNESS: No, we did not see a dependence steeper than V^2, although this could occur at higher voltages. We kept to lower voltages to avoid destruction of the units through excessive joule heating. Thus we cannot be sure that the observed dependence was a result of S.C.L.C. and as you say we cannot therefore deduce reliable mobility values.

J. H. BECKER: Did you check the thickness dependence of the current in the V^2 range? The data you showed did not seem to show a $1/d^3$ dependence. If it did not vary as $1/d^3$, what is the meaning of the mobility you calculated using V^2/d^3 formula?

CHAMPNESS: The measurements on Te–Se–Te structures were done with only two thicknesses of 14 and 22 microns. The two films had different crystallographic orientations and contained different nominal chlorine concentrations. Thus we could not look for a thickness dependence here. On the Cd–Se–Te structures we have not enough information yet to determine the thickness dependence either, but preliminary studies indicate that the current density increases more strongly than $1/d$. Possibly it may be nearer to a $1/d^2$ dependence. This does not, of course, correspond to the space charge limited current formula, but a $1/d^2$ dependence was observed by Graeffe and Heleskivi,[9] who explained their results using a model in which the low-resistance selenium was criss-crossed with high-resistance layers.

J. STUKE (University of Marburg): I would like to comment on the problem of space charge limited currents in selenium single crystals. We have some years ago also found a quadratic voltage dependence and a $1/d^2$ thickness dependence. The absence of a $1/d^3$ dependence does not necessarily mean that we have no space charge limited currents. In the case of space charge limited currents in internal barriers with a small thickness δ compared to the external thickness d, one obtains a $1/d^2$ relation. I think that this applies for trigonal selenium single crystals.

MAGNETOPHONON RESONANCE IN TELLURIUM

D. V. Mashovets and S. S. Shalyt
Semiconductor Institute, Academy of Sciences of the U.S.S.R., Leningrad, U.S.S.R.

It was observed in the investigation of the effect of a strong pulsed magnetic field on the electrical conductivity of non-degenerate samples of tellurium at 77°K that the plot of magnetoresistance against field intensity H is an oscillating curve with extrema that are periodic in reciprocal field $1/H$. Such a singularity observed in a non-degenerate semiconductor may be evidence of magnetophonon resonance, the phenomenon predicted by Gurevich and Firsov[1] and studied in detail in n-InSb.[2] In magnetophonon resonance the oscillation amplitude should decrease with decreasing temperature, and the period $\Delta(1/H)$ should be dependent on the carrier effective mass m^* and the frequency ω of the optical vibrations of the crystal:

$$\Delta\left(\frac{1}{H}\right) = \frac{e}{m^*\omega c} \qquad (1)$$

The following experimental data favour magnetophonon resonance as the cause of the observed oscillations in tellurium:

(a) The period of the oscillations does not depend on the carrier density p which ranges from 2×10^{14} cm^{-3} to 4×10^{16} cm^{-3}.
(b) The amplitude of the oscillations decreases markedly when the temperature is lowered from 77°K to 20°K and the optical vibrations of the crystal are damped.
(c) An estimate of the frequency from formula (1) yields a value close to the characteristic frequency of the optical vibrations for the crystal ($\omega \approx 10^{13}$ sec^{-1}).

The experimental curves (Fig. 1) pertain to two arrangements of the magnetic field H relative to the current j and main crystal axis C_3 ($j \| C_3$).

On the basis of considerations which have been detailed,[1] it should be assumed that the resonant values of H correspond to minima of the $(\Delta\rho/\rho_0)(H)$ curve when $H \| j$ and the maxima of the curve when $H \perp j$. The minima at $H = 47$ kOe, 70 kOe and 140 kOe show a periodicity in $(1/H) = 7 \times 10^{-6}$ Oe^{-1},

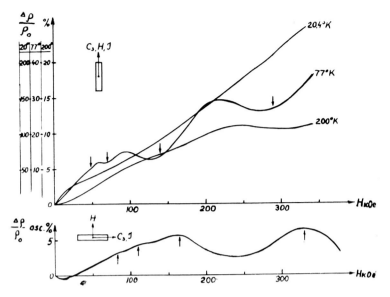

FIG. 1. Experimental curves showing the magnetoresistance dependence of tellurium single crystal on field intensity. Hole density 1×10^{15} cm^{-3}; mobility 8000 cm^2 V^{-1} sec^{-1}. Dimensions of the samples: $H \| C_3 - 2 \times 2 \times 20$ mm^3; $H \perp C_3 - 1 \times 1 \times 8$ mm^3. When $H \perp j$, the weak oscillating part was separated in the form of a difference signal by applying to the amplifier an out-of-phase signal which was proportional to the field.

and the maxima at $H = 85$ kOe, 110 kOe, 160 kOe and 330 kOe show a periodicity in $\Delta(1/H) = 3 \times 10^{-6}$ Oe^{-1}.

At present it is possible to give the following interpretation of the experimental results: taking into consideration the fact that the Fermi surface for the holes in tellurium has the shape of an ellipsoid of revolution, one obtains for the two cases investigated:

$$\Delta\left(\frac{1}{H}\right)_{H \| C_3} = \Delta_\| = \frac{e}{m_1 \omega_\| c}$$

$$\Delta\left(\frac{1}{H}\right)_{H \perp C_3} = \Delta_\perp = \frac{e}{\sqrt{(m_1 m_3)} \omega_\perp c}$$

These expressions enable us either to determine the frequencies of optical vibrations in the crystal if m_1 and m_3 are known, or to solve the inverse problem. In the first case it is possible by purely electrical measurements to determine the frequencies of the vibrational spectrum of a crystal. A quantitative analysis of the experimental curves can hardly lead at present to unambiguous results, since the available data on the physical properties of tellurium do not give reliable data either on the components of effective masses m_1 and m_3 at the temperature of the experiment or on the complete vibrational spectrum of the complex tellurium lattice. The experimental results may be interpreted at present as follows: reliable data for the values of m_1 and m_3 were obtained from a study of a sample with a very small hole concentration ($p \approx 10^{13}$ cm^{-3}) at $T=1.3°$K ($m_1=0.12\,m_o$, $m_3=0.25\,m_o$),[3] but from thermoelectric experiments with tellurium it is known that these values increase notably with increase of temperature. Since this dependence has not been established with sufficient accuracy, we do not see a possibility at present of evaluating the components at the temperature used in the present investigation of magnetophonon resonance ($T=77°$K). However, if it is assumed that the ratio $m_3/m_1=2.1$ remains unchanged with increase of temperature up to 77°K, then from the experimentally obtained ratio of the periods $\Delta_\parallel/\Delta_\perp=2.3$ we find $\omega_\perp/\omega_\parallel=1.6$.

From the whole complex of optical and neutron scattering investigations and theoretical considerations[4-7] it may be concluded that among possible optical modes in the vibrational spectrum of tellurium there are two with frequencies $\omega_1=1.7$ to 1.8×10^{13} sec^{-1} and $\omega_2=2.7\times10^{13}$ sec^{-1}. The ratio of these frequencies, $\omega_2/\omega_1=1.6$, agrees well with our experimental ratio, $\omega_\perp/\omega_\parallel=1.6$. To correlate absolute values for the frequencies $\omega_1=\omega_\parallel$ and $\omega_2=\omega_\perp$, it must be assumed that both components of the hole effective mass, m_1 and m_3, should yield a similar 25 per cent increase with rise of temperature from 1.3°K to 77°K. Such change is quite admissible when regarded in the light of available information on this topic.

In the above analysis a possible dependence of the effective mass on the concentration has not been taken into account. The fact that the hole concentration in the investigated tellurium samples did not exceed 4×10^{16} cm^{-3} may support this assumption.

The additional deep minimum (Fig. 1) at $H=290$ kOe, which violates the periodicity of equation (1), may be connected with the spin splitting of the Landau levels. However, the numerical value of the g-factor which can be determined in this case from the formula $g\mu_B H=\hbar\omega$ turns out to be too high ($g\approx 3$) compared with the results of paramagnetic resonance investigations of tellurium.[8]

The basic qualitative result of this paper is that in tellurium for $T>30-50°$K the optical vibrations of the lattice play a notable role in the scattering of current carriers. Such a conclusion perhaps may explain one experimental observation which has so far been incomprehensible. As has been shown,[9] in the purest tellurium samples with comparatively weak impurity scattering, the mobility increases in the region 15–50°K more rapidly than the law $\mu \propto T^{-\frac{3}{2}}$. In accordance with the above conclusion, such an increase may be explained by a freezing-out of optical phonons, thereby lowering more rapidly the intensity of thermal scattering.

REFERENCES

1. GUREVICH, V. L. and FIRSOV, JU. A., *J. Exper. Teor. Fiz.* **40**, 199 (1961); **41**, 512 (1961).
2. PARFEN'EV, R. V., SHALYT, S. S. and MUZHDABA, V. M., *J. Exper. Teor. Fiz.* **47**, 444 (1964).
3. MENDUM, I. H. and DEXTER, R. N., *Bull. Am. Phys. Soc.* **9**, 632 (1964).
4. KOTOV, B. A., OKUNEVA, N. M. and SHAH-BUDAGOV, A. L., *Fiz. Tverd. Tela* **9**, 2553 (1967).
5. LUCOVSKY, G., *Bull. Am. Phys. Soc.* **12**, 102 (1967).
6. GROSSE, P., LUTZ, M. and RICHTER, W., *Solid State Comm.* **5**, 99 (1967).
7. CALWELL, R. S., FAN, H. Y., *Phys. Rev.* **114**, 664 (1959).
8. HULIN, M., *Ann. Phys.* **8**, 647 (1963).
9. DATARS, W. R., FISHER, G. and EASTMAN, P. C., *Can. J. Phys.* **41**, 178 (1963).
10. FARBSTEIN, I. I., POGARSKY, A. M. and SHALYT, S. S., *Fiz. Tverd. Tela* **7**, 2383 (1965).

AUTHOR INDEX

Abdullayev, G. B. 179, 321
Adams, J. E. 293
Alpert, Y. 31
Asadov, Y. G. 179
Axmann, A. 299

Champness, C. H. 349
Cherin, P. 223

Drilhon, G. 31

Fitton, B. 163

Geick, R. 277, 309
Gissler, W. 299
Gobrecht, H. 87
Griffiths, C. H. 135, 163, 349
Grosse, P. 309
Guthmann, C. 47

Haas, W. 293
Haisty, R. W. 345
Hardy, D. 53
Harrison, D. E. 115
Harrison, J. D. 115
Henrion, W. 75

Ibragimov, N. I. 321
Iizima, S. 199

Keezer, R. C. 103
Kolb, E. D. 155, 213
Krambeck, R. M. 59
Krebs, H. 345

Lanyon, H. P. D. 59, 205

Laudise, R. A. 213
Lucovsky, G. 255

Madelung, O. 23
Mamedov, K. P. 179
Mamedov, Sh. V. 321
Mashovets, D. V. 371
Mooradian, A. 269

Nicolet, M.-A. 199

Queisser, H. J. 289

Richter, W. 309
Rigaux, C. 31, 53

Salo, T. 335
Sandrock, R. 69
Sang, H. 135, 349
Schröder, U. 277
Shalyt, S. S. 371
Springer, T. 299
Stubb, T. 335
Stuke, J. 3
Suosara, E. 335

Tausend, A. 233
Taynai, J. 199
Thuillier, J. M. 47
Treusch, J. 23

Unger, P. 223

Vaško, A. 241

Wright, G. B. 269

SUBJECT INDEX

Absorption edge
    selenium   75–77, 194–195, 249
    tellurium   33–34

Band structure
    selenium   10–14, 23–30, 59–83
    selenium pseudopotential calculation   69–74
    tellurium   10–14, 23–58
Birefringence, trigonal selenium   293–297
Bonding
    selenium   3–4
    tellurium   3–4
Brillouin zone, selenium and tellurium   23–24

Capacitance
    potassium-doped selenium   64–66
    selenium   59–68
    thallium-doped selenium   61–66
Charge transfer state in selenium   321–344
Conductivity, electrical
    selenium   335–347
    selenium–antimony–germanium glasses   345–347
Conductivity, thermal, selenium and tellurium   5
Crystallization behavior, selenium   163–167, 205–211
    effect of impurities   93–95, 106–112, 206–207, 210
Crystallization rate selenium by differential thermal analysis   206–207
Crystallization trigonal selenium from monoclinic form   184–194

Debye temperature, selenium and tellurium   5
Determination oxygen in selenium by neutron activation analysis   87–89
Dielectric constant, selenium   343
Differential thermal analysis, selenium   96–97, 206–207
Diffusion sulfur in selenium   90–92
Donor–acceptor interactions in selenium   321–334

Elastic constants, selenium and tellurium   3
Electronic paramagnetic resonance absorption in selenium   321–334
Electro-optic effect in trigonal selenium   293–297
Electroreflectance, selenium   99–101
Energy bands, selenium   71–72

Fermi surface, valence band tellurium   50–51

Hole concentration and mobility, tellurium   32
Hole mobility
    selenium   10
    tellurium   10, 32, 315–316

Impurities in selenium   15, 87–102
    effect on crystal growth   93–95, 106–108
    effect on physical properties   87–102
    effect on surface tension of the liquid selenium   98
Impurities in tellurium   15
Infrared absorption
    doped selenium   233–239, 250–253
    selenium   233–239, 250–267
    tellurium   221
Interband transition of holes in tellurium   53–58
Intraband transitions in selenium   82

Light beam modulation in trigonal selenium   293–297
Luminescence
    selenium   11, 289–292
    selenium–tellurium mixed crystals   289
    tellurium   11

Magnetoabsorption in tellurium   31–45, 54–55
Magnetophonon resonance in tellurium   371–374
Molecular structure, liquid selenium   103–106

Morphology and growth, trigonal selenium 163–177
Multiphonon absorption in selenium 82–83

Neutron activation analysis, selenium 87–89
Neutron scattering spectra
 selenium 17, 304–305
 tellurium 303–305

Optical activity, trigonal selenium 295–296
Optical properties, glassy vitreous selenium 241–254

Phonon energies, selenium and tellurium 6
Phonon scattering in tellurium 316
Phonon spectra, selenium and tellurium 8, 9, 277–287
Photoluminescence
 selenium 289–291
 selenium–tellurium mixed crystals 289
Piezoelectric effect, selenium and tellurium 8
Pockels constant, trigonal selenium 297

Raman spectra, trigonal, α-monoclinic and amorphous selenium 269–276
Reflectivity spectra
 selenium 7, 14, 77–81
 tellurium 7, 14, 311–312

Selenium
 absorption edge 75–77, 194–195, 249
 amorphous
  differential thermal analysis 96–97, 206–207
  infrared absorption 233–239, 260–267
  neutron scattering spectra 304–305
  optical properties 241–254
  Raman spectrum 274–276
  structure 255–267
 band structure 10–14, 23–30, 59–83
  pseudopotential calculation 69–74
 birefringence, trigonal 293–297
 bonding 3–4
 Brillouin zone 23–24
 capacitance 59–68
  potassium-doped 64–66
  thallium-doped 61–66
 charge transfer state in 321–344

conductivity
 electrical 335–347
 thermal 5
crystallization
 behavior 163–167, 205–211
 effect of impurities 93–95, 106–112, 206–207, 210
 rate by differential thermal analysis 206–207
 trigonal from monoclinic form 184–194
Debye temperature 5
determination oxygen in, by neutron activation analysis 87–89
dielectric constant 343
differential thermal analysis 96–97, 206–207
diffusion sulfur in 90–92
donor–acceptor interactions in 321–334
elastic constants 3
electronic paramagnetic resonance absorption in 321–334
electro-optic effect in trigonal 293–297
electroreflectance 99–101
energy bands 71–72
hole mobility 10
impurities in 15, 87–102
 effect on crystal growth 93–95, 106–108
 effect on physical properties 87–102
 effect on surface tension of liquid 98
infrared absorption 233–239, 250–267
 in doped 233–239, 250–253
intraband transitions in 82
light beam modulation in trigonal 293–297
luminescence 11, 289–292
 in selenium–tellurium mixed crystals 289
molecular structure of liquid 103–106
monoclinic 15–16, 179–203, 223–229
 bond distances and angles, α- and β-forms 183, 199–203, 226–229
 crystallization to trigonal form 184–194
 growth of α- and β-forms 180–183
 infrared absorption 259–261
 photoconductivity 194–197
 physical properties 194–197
 preparation and identification, α-form 199–203
 Raman spectrum, α-form 273–274
 single crystals 180–181, 199–203
 structure 183, 199–203, 225–229, 259–261
morphology and growth of trigonal 163–177

SUBJECT INDEX

Selenium—*continued*
  multiphonon absorption in 82–83
  neutron activation analysis 87–89
  neutron scattering spectra 17, 304–305
  optical activity of trigonal 295–296
  optical properties of glassy vitreous 241–254
  phonon energies 6
  phonon spectra 8–9, 277–287
  piezoelectric effect 8
  Pockels constant in trigonal 297
  Raman spectra, trigonal, α-monoclinic and amorphous 269–276
  reflectivity spectra 7, 14, 77–81
  single-crystal films 135–154, 349–370
  single crystals
    α- and β-monoclinic forms 180–181
    trigonal
      etch pit studies 115–134
      growth by epitaxy 135–154
      growth from aqueous sulfide solutions 155–161
      growth from impurity-doped melts 103–113
  solubility in aqueous sulfide solutions 155–158
  space charge limited currents in trigonal 336–338, 366–367
  structure
    α-monoclinic 259–261
    α- and β-monoclinic 183, 199–203, 225–229
    amorphous, from infrared measurements 255–267
    internal defects in glassy vitreous 241–248
    trigonal 223–224, 256–259
  superconductivity 15, 16
  transport mechanism 10
  trapping levels in 59–68
  trigonal
    band structure 10–14, 23–30, 59–83
    birefringence 293–297
    conductivity, electrical 335–344
    electro-optic effect in 293–297
    infrared absorption 256–259
    light beam modulation in 293–297
    luminescence 11, 289–292
    morphology and growth 93–95, 163–177
    optical activity 295–296
    phonon spectra 277–287
    Pockels constant 297
    Raman spectrum 269–273
    single crystals 95–96, 103–161
    space charge limited currents in 336–338, 366–367
    structure 223–224, 256–259
    vibrational modes 299–307
    viscosity 208–211
      effect of impurities 103–107
Selenium–antimony–germanium system, glass formation and electrical conductivity 345–347
Selenium rectifier, single-crystal type 349–370
Shubnikov–de Haas effect in tellurium 47–52
Single-crystal films, selenium 135–154, 349–370
Single-crystal selenium rectifier 349–370
Single crystals
  α- and β-monoclinic selenium 180–181
  trigonal selenium
    etch pit studies 115–134
    growth by epitaxy 135–154
    growth from aqueous sulfide solutions 155–161
    growth from impurity-doped melts 103–113
  trigonal tellurium, growth by Czochralski technique 213–222
Solubility, selenium in aqueous sulfide solutions 115–158
Space charge limited currents in trigonal selenium 336–338, 366–367
Structure
  α-monoclinic selenium 259–261
  α- and β-monoclinic selenium 183, 199–203, 225–229
  amorphous selenium from infrared measurements 255–267
  internal defects in glassy vitreous selenium 241–248
  trigonal selenium 223–224, 256–259
  trigonal tellurium 223–225
Superconductivity, selenium and tellurium 15, 16

Tellurium
  absorption edge 33–34
  band structure 10–14, 23–58
  bonding 3–4
  Brillouin zone 23–24
  conductivity, thermal 5
  Debye temperature 5
  elastic constants 3
  Fermi surface, valence band 50–51
  hole concentration 32
  hole mobility 10, 32, 315–316
  impurities in 15
  infrared absorption 221
  interband transition of holes in 53–58

Tellurium—*continued*
  luminescence 11
    selenium–tellurium mixed crystals 289
  magnetoabsorption in 31–45, 54–55
  magnetophonon resonance in 371–374
  neutron scattering spectra 303–305
  phonon energies 6
  phonon scattering in 316
  phonon spectra 8–9, 277–287
  piezoelectric effect 8
  reflectivity spectra 7, 14, 311–312
  Shubnikov–de Haas effect in 47–52
  single crystals, trigonal, growth by Czochralski technique 213–222
  structure, trigonal 223–225
  superconductivity 15, 16
  vibrational modes 299–318
  viscosity, selenium–tellurium alloys 208–211
Transport mechanism, selenium 10
Trapping levels in selenium 59–68

Vibrational modes
  selenium 299–307
  tellurium 299–307, 309–318
Viscosity
  selenium 208–211
    effect of impurities 103–107
  selenium–arsenic alloys 208–211
  selenium–tellurium alloys 208–211